So You Think You Can Live Without Oil

How dependent are you on Petroleum?

By

Robert Devine

Table of Contents

Dedication

This book is dedicated to all oil industry workers, from CEOs to roustabouts, producers large and small, to truckers, pipeline and refinery workers, lease buyers to traders. Industry Advocacy Associations and most especially to all the great people I have had the privilege of working with over the past 50 years.

Forward

This publication is a culmination of years of studying climate predictions made by seemingly endless sources, mostly government entities, politicians and master scientific Hollywood celebrities. The study looks at predictions from the 1950's through the year 2024. I am not a climate scientist but I am a student of climate studies. I have also worked in the oil industry for many years in the areas of refiner supply, transportation, trading and marketing. Therefore, I am familiar with the absolute need for petroleum products. Not just for motor fuels, and not just for the consumption here in the United States but Globally. To eliminate hydrocarbon products from the marketplace would return United States citizens and the world population to the 18th century and likely create a worldwide famine.

What makes this all electric movement so ludicrous is the fact that you cannot produce batteries for Electric Vehicles without Petroleum. That statement will be expounded upon in the pages to follow. Also, there is not enough infrastructure to support the proposed goals.

It is absolutely necessary for people to understand the need for petroleum. A few facts I will mention here is that the advent of petroleum products and the invention of the internal combustion engine were innovations that changed the course of history. The oil and gas industry quiet literally fueled the development of the United States. To point out some little known facts, one major accomplishment was to save the whales, (whale oil was the common source of oil for lamps) another was reducing pollution (manure production in cities across the country) and brought a prosperous economy to THE WORLD, not just the United States. As you read through make note of the things you use everyday that are Hydrocarbon based. Then using common sense try to suggest an alternative for those products.

There is a great deal of focus on the Biden Administration's war on the Oil and Gas industry. In addition, much of the focus in the pages to follow is on the state of California. California is the self -proclaimed leader of carbon reduction and in the forefront of climate change technology. California has introduced legislation to eliminate internal combustion engine (ICE) vehicles by 2035 where the only vehicles permitted to operate in the state will be electrically powered. While touting that they are the model of environmental stewardship they give not one thought to what is required to accomplish their goals.

In addition to California's misguided efforts to save the world by eliminating hydrocarbons, you see overzealous, ignorant younger generations taking to the streets carrying signs demanding we stop using petroleum. "STOP OIL" They glue themselves to walls in art galleries (with glue made from petroleum) and throw soup at art work or sit in the middle of busy streets blocking traffic. We are supposed to, immediately conform to their demands and appease these spoiled know nothing brats regardless of the consequences. They have no clue of what they are asking for. The fact is that most of the clothing they are wearing are petroleum based and those that are not were delivered to them using petroleum. Going "Electric" will not be a substitute for Petroleum! Electricity is a product but does not produce other products. You can't extract an ampere or a volt and make clothing, medicines, plastics or any other product that contains petroleum derivatives. Electricity is produced using hydrocarbons in one form or another. Solar panels and or wind turbines contain hydrocarbon elements that cannot be duplicated by any other source material. The only source for the hydrocarbons used in the production of panels and turbines must come from refined crude oil.

The oil industry has many facets from very large, fully integrated corporations such as ExxonMobil and Chevron who have global production as well as refining, transportation and retail assets. They are publicly traded and have stockholders that their Boards of Directors answer to. Then there are intermediate sized corporations/companies that are independents that are focused on exploration and producing oil with no downstream assets. Then there are the small mom and pop family owned producing companies. These companies, collectively, employ millions of people across the globe and produce trillions of dollars in tax revenues annually and millions of dollars of charitable funding.

The industry has several advocacy organizations that represent the different segments of the business from oil and gas exploration, production, transportation, refining to retail. One of the most known is the Independent Petroleum Association of America (IPAA), headquartered in Washington D.C. Their basic charter is to support the independent producers of America by advocating sound science legislation that encourages the energy security of the country and to attempt to discourage legislation that is detrimental to the industry. This explanation is an oversimplification of what they do but in the interest of brevity I'll leave it at that. The association has a publication called "Energy In Depth" wherein they gather pertinent information, relative to the Oil and Gas Industry, that have been published by numerous news agencies across the globe. These articles contain information that pertains to issues affecting the Oil and Gas industry Nationwide, Internationally, U.S Regional and State specific. They cover new projects and discoveries with positive and negative information on issues affecting the industry. In the pages that follow you will find several references to "Energy In Depth" articles. Many of the articles quoted refer to government interference and the negative impact on not only the industry but also consumers.

Take the time to read all of what is written. I think you will find it informative and enlightening and perhaps provide some sound science analysis to disprove the theories surrounding "Climate Change".

Chapter 1

How Dependent Are You on Petroleum

(Petroleum Not Just a Fuel and Nothing Moves Without a Truck)

Let's see if we can enlighten you. There are over 6,000 products that are commonly used by the general public that are petroleum based.

Take a look around your house and see if you can identify products containing petroleum. All plastics are petroleum based. Crude oil is processed in a refinery and produces not just gasoline, diesel fuel and jet fuel (motor fuels) but a long list of chemicals that are feedstocks for thousands of products you use every day. Have you ever given any thought to the products you use everyday that are petroleum based?

The petrochemical feedstocks mentioned are converted into a variety of products. Those petrochemicals are the basis for plastics used in manufacturing things used in your car: a variety of automotive components such as onboard computers, seats, floor mats, carpet, dash covers, instrument covers, steering wheels, upholstery, gear shift knobs, control buttons, front and rear bumpers (cars not trucks), tires, electronics, plug wires, engine coolant, lube oil, bearing grease, transmission fluid, power steering fluid, brake fluid, paint and many more products and components.

If we eliminate petroleum, what will we use to pave our new roads and driveways or repair older ones as asphalt would no longer be available. Asphalt is the bottom of a heavy barrel of oil, quite literally "tar". Will we use concrete for roads?? If so, how will the concrete get produced without oil? Concrete / cement is strip mined using heavy equipment fueled by diesel. In addition, the process requires a great deal of heat, one fuel used in the cement kilns is Petroleum Coke. This is a cinder-like product that comes from oil refinery "Coker drums". It is the extreme bottom of a barrel of heavy oil. The concrete manufacturer/miner will use the Coke in a kiln to dry limestone and other materials by laying down a bed of Coke and then spraying a diesel-like (but heavier) fuel over the Coke to assist in igniting it. Once ignited the coke will produce heat much like charcoal and will need occasional "top off" of coke as well as a possible shot of diesel fuel. The heat produced by the process produces what the industry refers to as clinker. The process of producing clinker produces a very high volume of CO2 which is also known as process emissions. Just how environmentally friendly is concrete as an alternative to asphalt. Not only does the process emit large volumes of CO2, you have to use a diesel fueled concrete truck to deliver the raw product to batch plants and then, once mixed, truck it to the road site and use additional petroleum fueled equipment to further construct the road. Living without petroleum puts us back to the early 1800's. Horse and buggy anyone?

Aside from the necessity for road construction, take a look around your home: you have cups, dishes, non-stick pans, televisions, DVR's, cell phones, landline telephones, calculators, earbuds, toys, tape and CD players, printer ink, Bluetooth equipment, musical instruments, household cleaners, blinds, shades and curtains, bicycle tires, bicycle parts such as seats, pedals, handlebar grips, and reflectors. Household items such as carpet and other types of flooring, home plumbing (water pipes), toilet mechanisms, shower heads. Other items found around the house such as plastic (rubber) gloves, furniture finishes, paint (even water-based paint has petroleum products), garden hoses, dishwasher detergents, plastic storage bags, plastic wraps, food storage containers, food packaging, plastic garden tools, bowl cleaning brush handles and bristles, trash bags, trash cans household and the commercial containers belonging to your waste management company. Many homes across the country have asphalt shingles and the use of "tar paper" is common in roofing. Many commercial buildings use roofing tar to seal the roof to protect the building from the elements. All are hydrocarbon based.

In Addition, personal products such as soap, toothpaste, shampoo and conditioners, hairspray and other hair care products, chapstick, lipstick, perfume, deodorant, makeup foundations, eye liner, mascara, hand lotion, shaving cream, disposable razors, and many more.

A variety of clothing from T shirts, dresses, blouses, suits, underclothes, yoga pants, shoes, work boots, hiking boots, ski boots, ski jackets, ski pants, running shoes, workout gear and much more. Sporting goods such as basketballs, footballs, soccer balls, golf balls, golf bags, golf club handles and graphite shafts, baseballs, astroturf, hockey pucks, skateboard and roller skate wheels, ice skate blade guards, boats, kayaks and canoes, surf boards, boat fenders, fishing rods, skis, snowboards, to name a few. Just about all clothing other than 100% cotton or wool is petroleum based. Cotton and wool items require petroleum to process and get the finished product to market. Petroleum based dry cleaning fluids are also required to maintain many types of clothing.

In the medical field you will find artificial limbs, heart valves, syringes, gloves, masks, face shields, goggles, gowns, IV components, catheters, salves, ointments, antihistamines, antibiotics, analgesics, cough syrups, suppositories, hand sanitizers, contact lenses, dentures, dental implants, hearing aids the list goes on. Medicines: aspirin, Tylenol, Ibuprofen, in fact 97% of all pharmaceuticals are petroleum based.

The above listings are just a small sampling of petroleum-based products. The challenge here is to try and get through one day without using a petroleum-based product. As mentioned above all of the products you use for everyday living are brought to you by petroleum in the form of commercial transportation. The exception might be if you are outback chopping firewood and even then, the steel in your axe would have been mined and processed using petroleum and the wood or plastic in the handle would have been produced using petroleum then put on a diesel powered truck at some point to bring it to market.

A post on Chevron USA's website Chevron.com, El Segundo refinery lists a breakdown of a single barrel of crude oil:

"What's in a Barrel of Oil"

"To some, a barrel of crude may look like a gooey liquid who's only redeeming virtue is to be eventually refined into gasoline."

"Researchers broke down a typical barrel of domestic crude oil into what may be produced. By the way, the average domestic crude oil has a gravity of 32 degrees and weighs 7.21 pounds per gallon."

The 32 gravity is a reference to the density of the oil. The gravity number is quoted based on the American Petroleum Institute or API. The institute converts specific gravity to "API" gravity using a mathematical table. To put it in perspective, oil at 13 API gravity will not pour at room temperature, it is much like tar, whereas a 32 API gravity will pour like water at room temperature.

"Here's what just one barrel of crude oil can produce:"

- "Enough liquefied gases (such as propane) to fill 12 small (14.1 ounce) cylinders for home, camping or workshop use.
- Enough gasoline to drive a medium-sized car (17 miles per gallon) over 280 miles.
- Asphalt to make about one gallon of tar for patching roofs or streets.
- Enough Lubricants to make about a quart of motor oil.
- Enough distillate fuel to drive a large truck (five miles per gallon) for almost 40 miles. If jet fuel fraction is included, that same truck can run nearly 50 miles.
- Nearly 70 kilowatt hours of electricity at a power plant generated by residual fuel.

- About four pounds of charcoal briquettes.
- Wax for 170 birthday candles or 27 wax crayons".

"There are enough petrochemicals left in that same barrel to provide the base for one of the following:"

"A petri dish with domestic crude oil being poured into it"

- 39 polyester shirts
- 750 pocket combs
- 540 toothbrushes
- 65 plastic dustpans
- 23 hula hoops
- 65 plastic drinking cups
- 195 one-cup measuring cups
- 11 plastic telephone housings
- 135 four-inch rubber balls

"The lighter materials in a barrel are used mainly for paint thinners and dry-cleaning solvents and they can make nearly a quart of one of these products. The miscellaneous fraction of what is left still contains enough by-products to be used in medicinal oils, still gas, road oil and plant condensates – a real industrial horn of plenty."

To further demonstrate the need for Petroleum

"Nothing Moves Without a Truck"

"Nothing Moves Without a Truck" is a slogan adopted by the Independent Trucking Association in California though I don't think it is exclusive as I have heard the same slogan repeated from other industries. Anything and everything you own, wear, eat, sleep on, sit on, live in, drive, all consumer goods you use and count on came to you on a truck. "NOTHING MOVES WITHOUT A TRUCK" Most items you own have been on a truck more than once. Starting with the raw materials, these materials are generally produced or gathered using diesel fueled equipment. Then those materials necessary to produce a product are then delivered to a manufacturing facility where the raw materials are converted into a finished product. Then the finished product will be loaded onto a truck for delivery to the buyer's facilities, usually a warehouse where items are sorted and assigned deliver locations. It will then, most likely, be moved from the warehouse by a truck, to a retailer. Electric vehicles (semi-trucks) cannot perform as efficiently as a diesel truck. More on diesel truck vs EV efficiency later in the chapter.

There are millions of trucking companies throughout the U.S. Some large commercial companies, some midsized, and many small independent companies. The small independent operators generally pick up small distribution jobs across the country. The independent operators are actually the backbone of the industry (and the supply chain). There are also specialty trucking companies which haul bulk products and materials that aren't directly used by the general public until further processing. These would be items such as construction materials like concrete, rock, wood, road oil, asphalt, water, and farm products (anything used or produced on a farm) such as milk, fresh produce, livestock, livestock feed, seed and a multitude of other products.

Overall, the profit margins for independents who move goods across the country are thin and competition is fierce. These operators can't afford the price of an electric truck and can't afford the down time to charge it when necessary. Forcing Electric Trucks on independent truckers will likely bankrupt the carrier. If you cut out the independent trucker from the supply chain the entire economy will suffer and many of the consumer products you are accustomed to being available to you will not be in stores. When and if they do show up the cost will be considerably more than what you are used to paying. Case in point would be the "Pandemic", more appropriately deemed the "planneddemic" as Anthony Fauci stated during a press conference after President Trump won the 2016 election "I believe President Trump will experience a Pandemic before the end of his term in office (don't believe it, Google it). The pandemic supply chain disruptions which were for a short period of time were due to truckers being confined to their homes therefore unable to deliver products to market. Elimination of diesel-powered trucks would have a much more severe impact (good-bye Teamsters Union).

The next time you are on a major highway take notice of the truck traffic. "Nothing Moves Without a Truck" is not just a slogan. The entire economy is dependent on moving goods from points of origin to the consumer and all stops in between. Let's take a look at the distribution system for just one product, a cotton shirt. No matter if it is a dress shirt, blouse or a "T" Shirt. The explanation below is somewhat condensed as the process of making fabric in a textile mill is more complex than what is identified below.

Consider all of the petroleum products necessary to produce the shirt. First you have to plant the cotton using diesel powered tractors to plow the field, plant the seed, cultivate the fields, fertilize the plants and then harvest the cotton. An aircraft is used to deliver (hydrocarbon based) defoliant to remove the leaves and expose the cotton fiber. Once harvested by a diesel powered "cotton picker" the raw cotton is transported to a cotton gin using diesel powered equipment. At the gin the seed is separated from the cotton fiber. Electric machinery is primarily used for this process. The seed is then moved by truck to plants that process it for replanting fields, livestock feed or pressed for cotton seed oil. Whatever the end product is, the raw seed will be loaded on a truck to a processing facility. Once finished it will be, once again, loaded on a truck and delivered to the market.

The cotton fiber is bailed at the gin and generally transported to a textile mill via train (diesel powered) where the raw fiber is transformed into yarn, thread and/or fabric bolts. The finished product of the mill is then transported by truck or possibly train to a facility where they produce products such as carpet, sheets, towels, clothing, and a number of different consumer products. Fabric bolts are put on a truck to a manufacturing facility where they produce the shirt. Then the shirt is back on a truck, most likely to a staging warehouse complex, then once again transported by truck to a retailer. Then it goes in your car to your home.

You can apply the same process to anything and everything in your home and or work place. Including your food supply. Take the truck out of the distribution system and the country goes hungry. Take note of the supply chain disruptions during the Pandemic. Grocery store shelves were vacant. Once again, truck drivers were stuck at home therefore there was no restocking the shelves.

The diesel-powered truck moves freely and efficiently around the country. There are truck stops at well- known locations which provide for efficient trip planning. It takes about 20 minutes to fuel a diesel rig which puts the driver back on the road quickly. The U.S. Department of Transportation restricts the hours of operation for drivers to 10 hours without an 8 consecutive hour rest period. The time to refuel is included in the operational 10-hour period if refueling is necessary during that time frame. Electric vehicles require a great deal of down time for charging and cannot handle the supply chain efficiently. The electric vehicle carries less payload due to battery weight, requires more down time to recharge and reduces the overall range of the truck. Payload, down time and range are all critical to the profit structure of the independent trucker.

Food distribution requires refrigerated trailers; Transportation Refrigeration Units (TRU's) also known as "reefer units", small light weight refrigeration unit. The TRU's are typically powered by the truck's main diesel engine while in transit to their destinations.

However, when the trailer is parked, loading, or unloading, the TRU can continue to be powered by the main engine or run off a separate standby engine. Where would the power come from for an Electric Truck? How much demand would be placed on the batteries? What would the additional load on the battery do to the range of the truck? What would run the TRU while charging the truck? Would the charge require additional time as the TRU would have to be powered by the same charger? Some of these units can run off of solar but those are not as reliable as the diesel engine units.

Next time you are at the grocery store take a look around at how many of the products you use require refrigeration. An electric tractor could probably cool a refrigerated trailer but for how long. An electric truck usually has a range of around 300 to 400 miles. The range is determined by the manufacturer using controlled environments that do not compare to actual road conditions (oval tracks). The range is dependent on payload and that is without the power drawn from a refrigerated trailer or even air conditioning, the vehicle headlights, running lights and does not consider traffic. Once the power is drained, I'm told that it will take as much as 3 hours to charge the vehicle, assuming the driver can find a charging station and there are no other trucks in line. The three hours would reduce their allotted 10 operational period to 7 hours. This would have a direct impact on efficiency and therefore profitability.

The Biden administration Federal Sustainability Plan proposed that federal agencies, (the largest fleet in the world), make 100% conversion to electric vehicles in the name of Climate Change (or Climate

Emergency???) by 2027. It also requires that 100% of federal medium and heavy-duty vehicles be converted by the year 2035. Your postal service may take a while. What of the vehicles required by the armed services, do these requirements put national security at risk? If the military vehicles remain petroleum powered instead of going electric, where will they get their fuel? It's clear that these rules are shot from the hip without any consideration for the impact on private parties, commercial entities, governmental agencies, or the nation as a whole, THE ENVIRONMENT is most certainly not a consideration. There is absolutely no mention of what is required to produce the equipment necessary to attain the proposed goals of an all-electric fleet or the economic impact on the country.

Further, according to an article in the Republican Investor "the EPA regulations apply to short haul and long-haul tractor-trailer trucks. In addition, the rulemaking will apply to vocational trucks like delivery vehicles, garbage trucks, school and public transit buses, concrete trucks and fire trucks. There go your Amazon, FEDEX, UPS, USPS and other delivery services rates, as well as postage stamp rates, they will triple at a minimum (how does $3 to mail a postcard sound). You think paying shipping is high today, just wait. You might also pray that you or your neighbors don't have a fire, depending on an electric vehicle the size of a fire truck is iffy at best. The truck might catch fire on the way to your house.

The agency (EPA) has projected that as a result of the RULEMAKING, by 2032, up to 50 percent of all new vocational trucks and 25 percent of new long-haul tractor-trailers will be electric. Once again, we have an out-of-control bureaucracy dictating "rules" to the industries without any proven justification. The government cannot tell us how much all of these requirements will help the environment and they won't admit how damaging it will truly be to both the environment and the economy. One 1,000 lb. car battery requires 500,000 lbs. of ore excavation in the mining process (for just the battery Cathode) then you can assume one 8,000 lb. truck battery will require 4,000,000 lbs. of ore extraction (more on this in a following chapter). What will be the total number of cars and trucks converted, multiply that by the numbers above. How can anyone say electric vehicles are environmentally friendly!

There are over 10 million trucking companies in the U.S. most of which are sole proprietors (Independent Mom and Pop operations) who depend on transporting goods for their livelihood. The margins for these truckers are thin because the competition is strong. The trucker's payload is what produces their profit. Many states have a maximum gross weight for the vehicle of 80,000 lbs. This includes the tractor, trailer and the goods being hauled. This is what is measured at truck scales you see along the nation's highways. The typical fully fueled truck and trailer will weigh between 23 and as much as 40 thousand pounds depending on how it is equipped. A tanker truck will be about 27 thousand lbs. while a truck with a flatbed or box trailer will be something less. Using a diesel tanker rig as an example at 27 thousand pounds allows for 53,000 pounds of payload to be highway compliant. An electric semi-truck will weigh about 8,000 to 10,000 pounds more than a diesel rig. The addition of batteries reduces the payload to about 43,000 lbs. at best. Reducing the payload reduces profit and increases cost. Which means everything that moves on a truck (and nothing moves without one) will cost the consumer more to make up for the lost payload. The reduced payload will also put pressure on the independent trucker and could very likely bankrupt the operator.

To exacerbate the situation, most independent operators cannot afford to pay $500,000 for a tractor and thin margins will not support the necessary debt to acquire the vehicle. This of course is compounded by the payload reduction. Also, the trucker would have to buy expensive charging equipment and pay not just the local kWh rate to charge the truck but that of enroute charging stations, those fees remain unknown and will vary state to state. That fact alone will make it difficult to determine trucking rates and still maintain a profit. Adding to the unknown is how long does it take to recharge a vehicle of that size? If a 100kWh vehicle (car) takes 30 minutes to charge what will the time be to charge a 900kWh vehicle? (more on that below). Also, a100kWh vehicle home charging system is expected to add the equivalent of one house to the power grid, what will a 900kWh vehicle add? Is the truck owner going to have to pay 10 times their current monthly electric bill? The cost of insurance is another unknown, it will likely be triple that of a diesel rig.

Therefore, forcing electric trucks on the industry will likely reduce the number of trucking companies and supply and demand will take over. Independent operators will either have to get a job with a larger commercial carrier or look for work in a different industry. With an approximate 10 million trucking companies in the country, how many drivers are going to be unemployed due to ludicrous “rules” set forth by bureaucrats. Also, the cost of goods will increase and the timing of delivery will increase. We are talking about severe supply chain disruptions. Your government at work. What happened to "by the people, for the people", the EPA is one thing that happened. Keep in mind there is NO VALID EVIDENCE to support these onerous "RULES".

Vehicle range will be reduced. A long-haul diesel truck has a range of up to 1,000 miles an electric truck will be about 400 miles or less, that would be a manufacturer's estimate. As mentioned, it is unknown what it would take to maintain a refrigerated truck, using an additional battery which would further reduce payload and therefore profit. An additional battery would also increase charging time. Factor in traffic, traffic accidents, the use of air conditioning, headlights and running lights which will all place additional load on the battery. Operating in extreme cold or hot conditions will have an impact. All of the above will reduce the projected range of an electric truck. The above mentioned 30 minute charging time for an electric car, give or take a few minutes is an estimate for a home charging unit. Using a commercial unit, it would depend on whether the charger owner/operator will allow a 100% charge or limit the charge to 80% of capacity. The truck battery will be about 900kWh which will take about 9 times the charging time of the 100kWh car battery. Charging an electric car, about 100kWh, you lose between 8% and 13% of the kWh in the charging system due to heat, what is the loss in a 900kWh charging system, is it the same percentage? Higher? You would be buying power that would never be accessible.

More increased cost to the operator, as well as increased cost to the consumer.

With the fact that electric vehicles have a reputation of catching fire, if I were an oil producer I would be a little nervous if an electric tanker truck showed up at my lease to load oil. I would be more nervous if I were stuck in a freeway traffic jam and stopped next to an electric tanker truck hauling petroleum products. Even a truck hauling a load of plastic toys would be a problem if it caught fire in a traffic jam. Toxic fumes from the vehicle batteries as well as the cargo would likely kill people in nearby vehicles.

There was an owner of a small trucking company (which will remain anonymous) in the Midwest that thought he would convert his fleet of about 30 trucks to EV's and phase out his diesel trucks. He approached the local policy makers in the town where his company operated to get approval and permits to install the necessary charging equipment for the conversion. The consensus was that it would take more power to charge his trucks than the entire town used and that the infrastructure was not in place to accommodate such a project. If the infrastructure were to be installed it would be at a cost to residents via taxes and increase in power rates and it would likely take years to accomplish due to the necessary environmental studies.

One thing to keep in mind is most of these so-called "Rule-Makers" have never held a private industry position. Most have been hired by government agencies base on their "academic achievements" or DEI which has absolutely nothing to do with the real world. Most depend on computer modeling to come to a "feasible" conclusion (garbage in, garbage out) regardless of the task at hand.

If this is allowed to continue, Russia, China and certain members of the UN will attain their goal of making the U.S. a third world country which they will control. What we need are politicians that are dedicated to the people and the country, not their campaign contributions, personal bank accounts and/ or inflated egos.

You can be certain that the radicals that are protesting petroleum are completely ignorant of the industry and what it provides. They have no idea of the products that are produced from petroleum. I don't mind the ignorance that can be fixed with the willingness to get educated, it's the stupidity that I find alarming, that can't be fixed. As we all are absolutely dependent on the petroleum industry in our everyday lives. Eliminating petroleum would put the country back by 175/200 years. Break out the candles; buy a horse; build a barn; plant a garden; chop some wood, kill some whales (you will need whale oil for your lamps). The petroleum industry put us in a position where we didn't have to sacrifice whales to service our reading lamps.

The advent of the internal combustion engine also reduced pollution, as prior to the introduction of the automobile, long distance land travel was supplied by steam locomotives which used wood or coal fired boilers to provide the steam. International travel was accomplished by ships using primarily coal for steam generation. We have all seen the thick black smoke generated by those boilers (western movies, and the Titanic). The primary means of localized transportation was supplied by the saddle horse and horse and buggy. That method of transportation came with a great deal of manure, much like Washington D.C. The birth of the petroleum industry actually created many other industries as power was fairly inexpensive and readily available.

Some politicians in our own government are trying to shut down the Oil and Gas industry believing they are saving the planet. Nothing could be further from the truth. Some of these politicians are Marxist, their motivation being control of the people, some are just going with the flow along party lines not knowing or caring about the consequences, others are just plain stupid, and all of them think you are stupid! Aside from the above mentioned necessity for oil and gas in your everyday life, you might consider what the industry contributes to the overall economy as it relates to jobs, tax revenues and charitable contributions.

An Article in Energy in Depth entitled **"USA Oil and Gas Supported Nearly 11 Million Jobs."** Rigzone –"The U.S. Oil and Gas industry supported nearly 11 million jobs and contributed nearly $1.8 TRILLION to the U.S. economy in 2021, according to a new study prepared by PricewaterhouseCoopers (PwC) which was commissioned by the American Petroleum Institute (API). PwC estimates that the U.S. oil and natural gas industry's operations directly or indirectly supported 9.4 million full and part-time jobs in the national economy in 2021, the study noted, adding that the industry's capital investment supported an additional 1.4 million jobs in the national economy. Combining the operational and capital investment impacts, the oil and natural gas industry's total employment impact on the national economy amounted to 10.8 full and part-time jobs in 2021, accounting for 5.4 percent of total U.S. employment, the study stated." This does not include the indirect jobs that are supported in the communities where these workers live. Businesses such as grocery stores, restaurants, retail outlets, convenience stores, medical facilities and any number of service businesses.

Another consideration might be, can you make your own clothes? Without a distribution system textile mills will be shut down as well as clothing manufacturers here in the U. S.

Items will likely be available from foreign countries but you may have to board a sailing ship to get there to buy things you need. A little bit of exaggeration but you get the idea. Other countries will not be doing what is being forced on U.S. citizens and industry.

We have identified issues associated with electric passenger vehicles as well as commercial trucking. What about Air Travel? Without petroleum there is no jet fuel or avgas, consequently, there will be NO COMMERCIAL or GENERAL AVIATION air travel. YOU HAVE BEEN GROUNDED!

No more flights for business or pleasure to anywhere you might need or want to go. How would anyone get to Hawaii? You would have to buy a sailboat. What happens to the tourist trade around the world and the associated economies when air travel is eliminated?

Air travel is all about weight and thrust. Do you think you will ever see an Electric Boeing 747, 757 or 777??? Replacing jet engines with electric batteries is an absolute impossibility. The weight alone would prevent the aircraft from flying. The size of the batteries necessary to replace the thrust (about 50,000 lbs.) from one a jet engine would likely be the size of a wing and weigh more than 300,000 lbs. That is just the battery, the wing structure to accommodate that weight as well as the stresses of inflight turbulence would be tons more weight. The plane would probably require two batteries. What would be the mineral requirements for a battery that size? You would be replacing jet engine thrust with propellers and a drive shaft from an extremely heavy battery system. How many props would be needed to replace one jet engine? The associated mechanical system to drive the prop would be another extremely heavy part of the aircraft. Where would you charge the batteries?? Pull over to "Cloud 9" and hook up??? How long would it take to charge an aircraft battery system of that size? What would be the range of the plane? Assuming you could get the plane off the ground the range would be extremely limited. There would be no overseas flights available. Charging would likely take days that would most certainly create scheduling issues. Where would the charging power come from? Keep in mind you would be dealing with hundreds of commercial aircraft. A jet aircraft is turned around and ready to be put back in service in a short time frame, maybe an hour or two. Further, battery systems do not perform well in cold climates. At altitude (30,000') you would experience temperatures in the neighborhood of minus 30 degrees. The cold conditions reduce the range of the system. How much reduction would not be readily apparent. Lithium -ion batteries produce power and abruptly stop when the charge is depleted, no gradual warning, like an old "D" cell flashlight! You cannot just pull over to a charging station and "gas up"! The fact that you could not get the plane off the ground in the first place renders all the above moot.

There have been some very light weight experimental aircraft (Gliders) that have used solar panel wing skins to provide power to a motor capable of turning a propeller for use when thermals give out but there is no way you could ever use solar panels on a commercial aircraft.

Battery powered Commercial Air Travel is well out into the future. there are some experimental projects on the drawing board one being a 90 passenger 8 propeller plane with a projected range of 500 miles. As mentioned there are a number of controlling factors that will determine the range of the aircraft. This concept will take a great deal of testing before the aircraft is put into production. A 500 mile range is not going to get you to Hawaii! In fact the practical range, to be safe, would likely be 300 miles.

Another news flash, you won't ever see electric general aviation aircraft. Using a Cessna 172 electrically powered as an example, the entire aircraft would have to be redesigned. Aviation is all about weight, balance, thrust, lift and useful load. The weight of the battery system replacing an aircraft engine would more than triple the weight of the Avgas fueled engine.

Using a newer Cessna 172 as an example, the aircraft, equipped with a 180 hp. Lycoming IO360 engine and a full load of fuel of 53 gallons weighs a little over 2500 lbs. The range is about 570 miles with FAA regulations requiring the pilot to hold a minimum of 30 minutes reserve of fuel on board. The useful load for the airplane is about 880 lbs. which includes passengers and luggage. If the airplane were to be battery powered what would the range be? What would the reserve be? Range will be impacted by weather conditions such as headwinds, storm systems, outside air temperature and altitude. The weight of the engine identified above is about 300 lbs. and sits forward of the firewall, the propeller and drive mechanism are mounted to the front of the engine. Weight and balance are calculated from the center of gravity aft of that point. Putting a 1,000 lb. battery in the same position would make the airplane extremely nose heavy. As a result, a controlled pattern descent for landing would be changed to a vertical descent (this of course assumes you could get off the ground). The electric battery to replace the engine would be at least a thousand pounds and very possibly more. Installing a battery as the primary source of power would shift center of gravity toward the aft end of the plane and would require a long "Drive Shaft" straight through the passenger compartment to turn the prop. This assumes you could carry passengers in such an airplane.

There is a small experimental electric airplane being built that looks to be a one place (Pilot only) configuration. The information given does not provide the airspeed capabilities or the useful load of the airplane. There will likely be extensive testing to determine the true numbers associated with those statistics.

Battery power, other than to start the avgas powered engine, in any form for aviation just won't work! And once again, how do you make the battery without petroleum?

We have grounded you with regard to aviation once we eliminate motor fuels, let's take a look at ocean voyages. Large ships use what is known as Heavy Fuel Oil (HFO) also known as "bunker fuel". Most engines for large ships have a dual fuel capacity where the primary fuel is HFO they also use distillates also known as Marine Diesel Fuel (MDF). Most ships will use the bunker fuel (HFO) for the long haul and switch to diesel when approaching a port. Air quality regulations require the ships to make the switch to diesel some miles off the coast of many U.S. Ports.

The list of global consumer goods brought to every country, not just the U.S., by the maritime industry is huge and not limited to products for the everyday consumer. There are a multitude of industrial products such as machinery for industries that supply other products for your needs. We can move domestic products without a ship but any and all exports would remain in the U.S. A total disruption of the trade balance would put the trade deficit in the trillions of dollars (this assumes only the U.S. is prohibited from using petroleum). Not too many people would consider the above when proposing the elimination of petroleum. Global shipping is not an issue recognized as being part of their everyday lives.

Let's take a look at something more common to the average citizen, the cruise ship industry. Many people in the U.S. look to cruise ships and their destinations as the ultimate vacation. News flash, cruise ships use the same bunker fuel and marine diesel as does the commercial shipping industry. Therefore, elimination of petroleum shuts down every cruise ship line globally. You will have to buy a sailboat!

To summarize: Without petroleum there are several thousand products that you use everyday that would no longer be available. There would be a major disruption in the supply chain as trucking companies go out of business and major shipping company rates (FEDEX, UPS type carriers) would likely triple just due to the cost and limited range of new equipment. All AIR TRAVEL would cease as well as the business and tourist economies around the world that are supported by air travel. The global shipping industry would be shut down which would likely create a famine in some countries and the elimination of the supply chain in just about every country. The cruise ship industry would be none existent as well as the tourist industry it supports. While there may be a possibility of an electric aircraft in the future it will not be built without PETROLEUM.

More reasons converting everything to battery power won't work are described in the following chapter.

Chapter 2

Electric Vehicles - What's Required?

The Mining Process and it's Environmental Impact

The government and environmental zealots have been pushing a narrative that Electric Vehicles (EV's) are the cleanest on the market and the best option for the environment. Let's take a look at how the batteries for those vehicles are produced.

The minerals required for ONE battery for an EV (Passenger cars not semitrucks). The makeup, a conservative estimate, which varies depending on the particular battery and application, is approximately 77 pounds of Nickel, 44 pounds of Manganese, 30 pounds of Cobalt and 17 pounds of Lithium, some batteries require as much as 165 lbs. of Lithium. Those minerals are used in ONLY the cathode part of the battery. Aluminum, iron, graphite, copper and steel makeup various other parts of the complete battery unit. In order to acquire (extract) the necessary minerals for JUST the cathode of ONE 1,000 lb. battery, it requires the excavation of approximately 500,000 pounds of ore (collectively).

These minerals are, for the most part, STRIP mined, meaning they are extracted from a large open pit using primarily diesel-powered equipment (google images of mines for any of the minerals listed). The U.S. has only one nickel mine which is an underground mine in

Michigan. While there are others being explored but not yet developed there is only one lithium mine located in Nevada. Therefore, to make batteries for electric cars here in the U.S most of the nickel and lithium must be imported from foreign countries. All of the Cobalt and Manganese must be imported from countries like China, South Africa, Argentina, Gabon, Brazil and others. There is a possibility of a manganese start-up mine in Arizona. The permitting and development time will likely take years. One thing to keep in mind is that Manganese is dangerously toxic to human organs.

Once extracted by diesel powered or electric powered "Dipper Shovels" which dig the ore from the pit, then load the ore onto an ore truck. The ore truck is a very large diesel-powered vehicle that uses as much as 92 gallons of diesel fuel per hour and they run 24/7. It's not miles per gallon; it's gallons per mile. This is common to most mines. In addition to powering all of the heavy equipment the mining facilities require diesel generators to power various other equipment such as conveyor belts, rock crushers, management facilities, lighting and charging equipment. Without PETROLEUM there is NO MINING!

These mines are located in primarily remote locations in countries around the world. This presents many logistical challenges in fueling these operations. There are no fuel pipelines to these mines therefore they have to rail or truck all the necessary diesel fuel to the site. It is likely the emissions levels from the equipment used to deliver fuel, extract minerals, transport the minerals, refine the ore, then construct one battery, exceeds the lifetime emissions of one internal combustion engine powered vehicle compared to battery life.

The processes for refining the minerals from the ore requires first that the ore be crushed to a fine powder. That is accomplished with large rock crushers, some electric, some diesel powered. Electrically driven crushers require diesel generators to supply the power. Using nickel as an example after crushing the powder is then roasted, smelted and converted to a matte, all using hydrocarbon energy. The crushed ore is then exposed to high heat that requires the use of hydrocarbon fueled furnaces. Extremely high temperatures of over 2600 degrees Fahrenheit are needed to liquify the metal. Sulfuric acid is also used in the process. The smelting process releases air emissions that are a contributing factor in air and water pollution. Sulfuric acid used in the process can cause acid rain, acid mist being formed from smelting plants and released into the atmosphere. Not a consideration in the political viewpoint when promoting "environmentally friendly" electric vehicles.

Nickel is a primary component of stainless steel. If the focus for nickel production is diverted to batteries, supply and demand will make all things stainless steel much more expensive. Many household appliances are made of stainless steel and about all commercial kitchens as well as food processing production/packaging facilities are stainless steel. Hospitals and other medical facilities use stainless steel for most of all of their operating room equipment so you should expect medical procedures to increase in price as well. The auto industry and many other industries rely on stainless steel equipment. Therefore, just about any product you buy will go up in price as the equipment to process it is more costly. An example: The impact on food processing will be evident as your grocery bill will go up with it. Commercial kitchen equipment and supplies cost go up so the price of dining out will go up as well.

Let's take a look at the other required minerals for the battery.

Manganese - There are no Manganese mines in the United States. All the manganese used in electric vehicles must be imported from China, Argentina, Brazil, Ukraine, South Africa, India and other countries around the world. In addition to the needs for Manganese in EV batteries, Manganese is an essential component of steel. In fact, there are no known substitutes for Manganese in the process of converting iron ore to steel. Therefore, if you divert manganese production to making EV batteries you will increase the cost of the mineral and consequently increase the cost of steel which is used in more necessary applications than can be listed here. There are manganese deposits in south east Arizona but to my knowledge they have not yet been developed into mining operations. With the necessary permitting and environmental studies, it will be years before the mine would be approved. If they go into production, they will be over-burden by EPA regulations causing the cost of extraction to increase considerably and likely shift the focus to foreign production. As mentioned earlier Manganese is very toxic to human organs so the EPA would be heavily involved in "Rulemaking" for any mining facility producing the mineral in the U.S.

Cobalt- There are no operating Cobalt mines in the U.S. Most of the Cobalt used in the manufacturing of Lithium-ion batteries comes from China or the Dominican Republic of the Congo. There was one Cobalt mine in the U.S in Idaho but it shut down before it started production. Mining faces a myriad of operating issues not the least of which is price of the product. Prices for Cobalt fell considerably in 2022 forcing the Idaho mine to shut down before it became fully operational. To further compound the pricing issue, the nearest Cobalt refining facility is located in Finland which adds a considerable cost to producing the finished product. The Cobalt would have to be transported to a seaport, loaded on a ship and taken to Finland where they would refine the ore and then load it back on a ship to bring the finished product back to the United States. The Idaho mine would have to load their ore on a train and ship it to the ports of either Seattle or Los Angeles to load it on an ocean going vessel. The trip to Finland from either of those ports would be long and costly. Plus, the fact that the train and all ocean going transportation uses hydrocarbon fuel. Once in the states the cobalt would be loaded back on a truck or a train and moved to a battery production facility. Much of the cobalt coming from the Congo region of Africa uses child labor to produce the ore.

Lithium – There is one operating lithium mine in the U.S in Nevada. The mine produces about 5,000 tons of Lithium per year which is about 2% of the world supply. Most of the U.S Lithium deposits are located in Nevada and Oregon. Lithium is imported to the U.S from mostly Australia, Chile, China and Argentina. Can anyone say Trade imbalance?

Lithium is by far the most volatile mineral used in making batteries, it is very flammable and not very stable. Batteries that short circuit due to accidents or just poor design will burst into flame due to lithium. The fires caused by lithium are extremely difficult to extinguish and are very toxic requiring fire fighters to wear breathing equipment. Equipment that would not be available to the general public if they were exposed to the fumes from an electric vehicle fire. Keep in mind it is not just the lithium burning but the collective minerals used in the various battery components, including the highly toxic manganese.

Copper- It requires about 132lbs. of copper to make one EV battery, while the above minerals are used in the cathode section of the battery, copper is an integral part of the anode. Further, one offshore wind turbine requires 4.7 tons of copper for power generation. The cooper is used in the generator windings and necessary internal wiring as well as transmission cables. (more on wind turbines in a later chapter)

An article in Fortune magazine states "that we don't have enough copper to make this political pipe dream of EV conversion in the next 10 years. There are not enough copper mines / production in the world today." The article further states that it takes "23 years to open a single copper mine". Assuming the reserves are economic enough to warrant excavation.

While the U.S. is currently an exporter of copper having about 11 copper mines, they are not capable of supplying the necessary volume of copper for the manufacture of the mandated electric vehicles not to mention wind turbines or the necessary infrastructure to support the EV's. Therefore, all of the above imported minerals, to include supplemental volumes of copper would be mined and brought to the U.S. on vessels powered by hydrocarbon fuels. Many of the minerals are shipped in the raw state of Ore. Once arriving at a U.S port the ore is then loaded on trucks and/ or trains and moved to stateside destinations for processing. Copper ore would arrive as gravel or golf ball sized "nuggets". This material would require additional crushing to make the ore a powder which would then be heated using hydrocarbon fueled furnaces in order to liquify the copper. Once the metal is rendered into a usable form for the market it would then be loaded back on a truck to a battery manufacturing facility. Once again, we experience emissions from the hydrocarbons in the effort to make a "supposedly" low emissions vehicle.

One scenario relative to the Biden administration's "surrender" of Afghanistan to the Taliban was due to pressure from the Chinese. Under the Trump (45) administration there was a plan for an orderly troop withdrawal with negotiated terms that the Taliban had accepted (I believe they were somewhat intimidated by Trump's offer). The withdrawal was planned for May of 2021. The negotiations did not include leaving 85 BILLION dollars' worth of arms and equipment nor did it include leaving behind American citizens and Afghan citizens that assisted in U.S operations over the years. When the announcement was made that the U.S would be pulling out of Afghanistan the Chinese government moved in to negotiate mineral rights agreements to gain access to mine the vast amounts of natural resources in Afghanistan which include lithium and copper. When we did not pull out in May of 2021 the Chinese became impatient with Washington and demanded that we leave so they could move in and commence their mining operations. Biden's fear of China (or what they might disclose relative to him and his family) put him in a position to prove his absolute incompetence.

An article in a google search under the title "Who is mining lithium in Afghanistan" you will find _ " Chinese investment in Afghanistan's lithium sector, A long shot in the short term. Speculation is mounting that China will take advantage of the power vacuum by the 2021 U.S. withdrawal from Afghanistan and seek dominance over that country's mineral resources, particularly its lithium deposits."

How feasible is it to expect the conversion of passenger vehicles to Electric power?

In 2022, the state of California alone had approximately 14.2 million vehicles registered; an estimated 5,000 were EV's. If you consider the requirements to convert the remaining 14 plus million registered vehicles the mineral requirements are astronomical. Example: 77 pounds of Nickel per battery would be over 1 billion 93 million pounds, just for California. The governor, in September of '22 requested that those people owning EV's not charge them as it would put excess demand on the grid and cause brown and black outs across the state. California did not have the power infrastructure to maintain the 5,000 vehicles they had then, where will the power come from for the remaining 14+ million cars? What would be the mineral and power requirements for converting all 283 million U.S registered vehicles?

At California's average of between $.33 and $.39 per kilowatt hour currently, an electric car is a luxury. When the number of cars increases so will the power cost to charge them. You can expect electricity rates to soar, the rates would likely exceed the cost of gasoline even considering California current high prices at the pump. The charging rates at commercial chargers are likely already higher than gasoline prices. There is no way to separate electric vehicle charging power from residential and industrial consumption. So, even if you don't own an electric vehicle, you will be paying extremely high rates per kWh for your home and products from industrial facilities will go up in price as well.

The intent of these onerous rules and laws is to ultimately eliminate privately owned vehicles and force the population into public transportation. Much like China, ride your bicycle or pay the state to ride their bus. Look at what would be required under those circumstances in order for you to maintain your current lifestyle especially as it pertains to your mobility. More of your rights are abolished!

Looking at the cost of electricity while the national average is $.16.7 per kWh California's power costs, with the push to eliminate fossil fuels, have surged to record highs.(noted above). News flash, every grocery store uses a great deal of electrical power for lighting, freezers, and refrigeration of perishable products. As electricity rates increase, so will your food costs.

An article in Energy In Depth (IPAA) demonstrates what is transpiring relative to the "Carbon Free" movement and the impact on the California electricity market.

California's 'ambitious' policies to reduce greenhouse gases drive a surge in electricity costs. Fox Business. Electricity prices in California have surged significantly over the past several years, which officials say is partly due to green energy policies being pushed throughout the state. The Legislative Analyst Office (LAO) released a new report revealing that monthly electricity rates for residential customers in California have increased by an average of nearly 50% since 2019. "In general, average residential electricity rates in California have grown faster than inflation in recent years, rising by about 47% over the four-year period from 2019 through 2023, compared to overall growth in prices of about 18%," the report reads.

People buying electric vehicles think that they are saving money because they are not exposed to the prices at the pump. They give no consideration to the cost of electricity, not just to charge their vehicle but to maintain their household. The increased demand raises the rates for all electric power, not just charging equipment.

Consider the cost of an EV relative to an internal combustion engine vehicle. The cost is usually about 40% higher (not including government subsidies, your tax dollars at work). Then you purchase an in-home charging system, with the cost of equipment and installation, at about $10,000. It is said that the installation of an EV charging system is the equivalent of adding another house to the power grid. Therefore, you can expect your electric utility bill to at least double.

Making enough electric vehicles to satisfy the needs of the American people is an absolute impossibility (283 million vehicles). Recent studies show there are not enough copper mines in the world to provide the materials needed to supply the number of electric vehicles and wind turbines for U.S. consumption. There are not enough minerals on earth to accomplish such a pipe dream. Nor are there enough mining companies to extract the minerals. If you think the cost of an EV is high now wait and watch as the minerals become scarce. What will a replacement battery cost?

Once replaced with a new one, the old battery will need to be disposed of. What is the cost of disposal? Some parts are recyclable but not all which means landfills. States charge a disposal fee for used tires by imposing a disposal tax, the lowest I found is Indiana at $.25 per tire, the highest Oklahoma at $5 per tire. If they charge these disposal fees for tires, what will they charge for Battery parts? Right now, only 5% of EV batteries are recyclable. The rest needs to be disposed of and you can rest assured that the party who owns the battery will be responsible for the disposal costs, as well as the replacement costs.

Mentioned above is the need for hydrocarbon fuels to produce the minerals for EV battery construction. What of the rest of the vehicle? It is made of, for the most part, plastic which is a hydrocarbon-based product. What are the alternative products for that application? Will we see the return of the WOODY?

In addition, all of the minerals for the batteries have to be processed and transported using hydrocarbon fuel (primarily diesel fuel), not to mention the chemicals necessary to separate the metals from the ore. The majority of the chemicals are hydrocarbon based.

The objective appears to be to eliminate hydrocarbons as an energy source but hydrocarbons are not just a fuel (as mentioned in the previous chapter). Where will the energy come from to mine, process and transport all the minerals necessary to make a battery? Where will the energy come from to generate the necessary electricity to charge batteries?

Looking at the scope of the project using California requirements alone. As pointed out above the state has 14.2 million registered passenger vehicles. To make that many batteries, as stated above, would require about 1.1 billion lbs. of Nickel. You can do the math on the rest of the minerals. (44lbs. Manganese, 30lbs. cobalt, 17 lbs. Lithium x 14.2 million). Keep in mind it takes 500,000 lbs. of ore, collectively, to make one cathode for one battery. Oh! I am so proud to be saving the planet! You will also need the 132 lbs. of copper along with an unknown amount of steel and other metals for the anode part of the battery. You then have to wrap the battery in something to keep all the components contained. You can use plastic, oh wait! Plastic in a petroleum product. Maybe burlap would work.

What power grid would be necessary to charge all of those vehicles? How much diesel fuel would be required to extract, process and transport that much raw material? Considering the above information, just how environmentally friendly is an electric vehicle?

Pacific Gas & Electric in California now wants to access electric cars batteries, through home charging stations, to draw power back into the grid in peak demand periods. So, they will bill you based on the energy used to charge your battery but if they need the power, they will suck it out of your car, chances are they won't pay for it (no metering system) but if they do it will be at some reduced rate relative to what you paid to charge your car. Further, you will not know when they draw your battery down and you will be out of luck when you attempt to drive to work in the morning. This whole electric conversion effort will put the U.S. back about 200 years. Buy a horse while there cheap! Talk about pollution, just envision everyone in New York City owning a horse. Pollution, by the way, was the reason for inventing the internal combustion engine in the first place, no place to put the horse manure. New York used to barge the manure out to the ocean to dispose of it.

China, India, Russia and the rest of the world are not complying with the mandated emission reductions that are being imposed on U.S citizens and industry. China, for example, in 2022 was building COAL FIRED power plants at the rate of one per week, from what I understand is that rate has now doubled.

Go online and google any mineral you want to see how it is mined. You will find for the most part they come from strip mines (pit mines). Pull up a picture of a nickel, copper or manganese mine and see if you think it is environmentally friendly.

Safety and Additional Costs

What percentage of the American people can afford an electric vehicle?? It is not just the cost of the vehicle it is a myriad of associated expenses. Insurance on the vehicle, home insurance, home charging unit, increased electric bills, replacing tires. EV's are heavier than standard internal combustion engine vehicles so they wear tires out at an accelerated rate (about three times the rate of Internal combustion engine vehicles).

Electric vehicles are prone to catching fire. In Boulder, Co. a family had an electric car in their garage that started smoking. They evacuated the house before the car burst into flames and called 911. The fire department put out the fire before the house could catch fire. This happened during the day, imagine if it happened at night when everyone in the house was sleeping. This is not an isolated incident; electric cars have been bursting into flames all across the country. We have all seen the video on the news of the electric bus going up in flames (the Biden administration wanted to put your kids in electric school buses).Brings to mind Kamala's love of yellow school buses. Try to imagine what would happen if an electric tractor-trailer (18-wheeler) hauling toxic materials were to catch fire? What would happen if that vehicle were to catch fire while stuck in a traffic jam. The situation would cause the fire department great difficulty in getting on scene. A load of plastic toys would be toxic in a truck fire. What would you do if you were stopped next to a burning electric truck in a traffic jam??? What if the truck was carrying chemicals or fuel?? How many would die??

The weight of an electric car is considerably more than that of an equivalent gasoline powered vehicle, as much as three thousand pounds heavier. This is a concern when parking in multi-level garages. There have been cases where floors have collapsed due to too many electric vehicles parked on the same level. The collective weight of the cars exceeded the design capability of the structure (2023 New York City). It is also a concern when buying tires. As mentioned above, the extra weight of an EV will wear out tires much faster than a gas powered vehicle. The tire wear is also more of a pollutant than tailpipe emissions. Therefore, your environmentally friendly Electric vehicle is not so friendly.

The insurance on electric vehicles is already on the rise as insurance companies discover that electric vehicles are so much more costly to repair if they can be repaired at all. Batteries that are in any way damaged, have to be replaced, there is no way to repair them. The cost of replacement can be as much as $20,000 plus the cost of the old battery disposal. Another issue for insurance companies is that the risk of personal injury in an accident is increased exponentially by the weight of the vehicle. This would apply to all who are involved in the accident not just the EV owner/driver.

One problem with raising insurance costs for EV's is that the rate needed to protect the insurance company would be very high and EV owners would likely not be able to afford it or be unwilling to pay the rate. Therefore, insurance companies are spreading the financial burden. They are increasing the cost of gas powered car insurance to cover what they can't get from EV owners. All in the interest of making sure the EV's will still be on the road supposedly preventing "Climate Change". So once again you are paying to subsidize EV owners.

Insurance companies are also looking at home charging equipment. If it hasn't happened yet you will likely see homeowners' insurance rates take a substantial increase for those having charging equipment installed. If you reside in an HOA community it is likely that the association will require a separate policy for the electric vehicle and the charging system naming the association as additional insured due to the high risk of fire. This may possibly happen without the charging equipment installed, just the fact that there is an electric vehicle parked in the garage. Lithium-ion batteries are very unstable and can short out causing excess heat and fires.

If you live in a condo complex and one of the residents owns an electric vehicle you are exposed to the possibility of that car catching fire and taking out the entire complex. If your insurance company is made aware of the EV exposure your home owner's insurance will likely increase.

If a gas-powered vehicle catches fire it will burn at approximately 1,500 degrees Fahrenheit and require about 2,000 gallons of water to extinguish. If an electric vehicle catches fire it will burn at 5,000 degrees Fahrenheit and require about 20,000 plus gallons of water to extinguish while releasing far more toxic fumes than the gas-powered vehicle fire.

According to testimony in Congressional hearings extinguishing lithium-ion battery fires is very difficult. Case in point a Telsa caught fire in a Houston suburb after an accident and it took eight firefighters an excess of 7 hours and 28,000 gallons of water to extinguish the fire. Will the cost of public services go up via taxes due to the exposure to electric vehicles?

https://www.kcra.com/article/placer-county-big-rig-fire-crash-interstate-80-emigrant-gap/61912862

In California, north and east of Sacramento, a Tesla 18 wheeler caught fire on Interstate 80 and completely burnt to the ground while shutting down the freeway in both directions for hours. It took the fire department 16 hours to extinguish the fire. It's been said that the toxic fumes from the fire could be deadly, fortunately the truck was not in a populated area but did raise concerns of a forest fire. This EV fire has prompted state legislators to request the governor to delay his EV mandate for electric trucks until more efficient means of extinguishing EV fires can be developed (What a concept!). Of concern is what happens if a similar incident occurred in a largely populated area such as Los Angeles. The toxic fumes from battery fires are a real health issue in areas of high population. This is another example of dangerous government policies being forced on the American citizen before the consequences are known. If you were stopped in traffic due to this accident and in your car sitting downwind of the fire, what is the possibility of inhaling toxic chemicals from the fire?? Would the fumes be deadly?

The cost of the installation of a public charging station is about $15,000 each when considering the necessary equipment and the utility company connection fees. The owner of the charger must make a profit in order to keep the equipment operational. Some studies claim that the national average cost per kilowatt hour (kWh) is about $.16.54. That may be so for some home chargers depending on their location. California will be $.28 to $.42 per kWh. Charging using commercial equipment will cost more as the charger owner will have management, upkeep and utility fees. In addition, the owner will have to make a profit. I saw a photograph of a commercial charging station that showed the kWh fee of $.49. According to the AAA the national average cost per kWh at commercial charging stations is $0.34.2. The total cost of a charge would depend on the size of your vehicle battery. You must also consider the 8% to13% loss of power due to heat generated in the process of charging for which you are paying. EV owners will tout "I save so much because I don't have to stop at the gas station" but they don't say what it costs them to charge the car. They will also claim that they are saving the planet due to less carbon emissions. The tire wear on an EV is about 18 times the level of tailpipe emissions of a comparable gasoline powered vehicle. They also discount all of the emissions necessary to make the battery.

The Federal government and state governments rely on taxes they impose on gasoline and diesel fuel sales. Federal gasoline tax is $.18.4 per gallon; diesel fuel is $.22 per gallon. The state taxes vary by state, California being the among the highest in the nation at $.59.6 on gasoline and $.45.4 on diesel. In addition, California adds a 2.25% sales tax based on the price at the pump which is a variable depending on market demand and location. Other states will also have a sales tax that will apply to the price at the pump. In California the prices include the addition of the carbon tax via the Cap-and-Trade program placed on the refiners in the state which can be very high and are variable. These fees are passed on to the consumer in the wholesale price of the product which means it is not readily apparent to the consumer, It is part of the pump price. The fees are imposed by the state however, the state government will classify it as corporate price gouging. Those greedy oil companies.

The more EV's that enter the market the higher the strain on state and federal income with the loss of the gasoline/diesel tax revenues. Therefore, you will see both the federal and state governments proposing a mileage tax on your travel. You will likely have to report your mileage on your tax returns to the IRS and your state tax board. This is a very desirable scenario for the government, they would like nothing better than to track your travel. Once established it would be relatively easy to legislate limits on your travel. There would have to be a new government agency, state and federal in order to monitor your activities. Larger government, how nice. If there is no state income tax in the state where you reside, they will have to form a new department likely within the DMV to assess your tax fee for your annual travel. The cost per mile is unknown, the federal number will be fixed but state fees will vary state to state. Expect those states with the highest fuel tax to be the highest per mile fee. God help you if your odometer stops working, they will estimate your mileage for you. Paying those fees at the pump each time you fill your gas tank is one thing, imagine if you were to get a tax bill once per year based on your miles traveled. The fee would likely be an estimate charged monthly and settled up at the end of the year much like income tax. Better ad a new line item to the budget! Initially the fee will apply to EV owners but once they get a taste of the revenues and the added power it is likely they will include hydrocarbon powered vehicles as well. Chances are they won't reduce the pump taxes in the process.

The current electrical grid in the U.S. is old and not equipped to handle the influx of increased demand placed on it by electric vehicles (add one house per vehicle). Many states are experiencing brown and black outs due to excessive demand on the system, California and Texas to name just two. As mentioned earlier the governor of California in September of 2022 made a request of EV owners not to charge their vehicles as it would put a strain on the power grid and create power outages across the state. What has happened is the federal governments under the Biden administration and state governments put the proverbial "cart before the horse". The objective is to force the public to move to electric vehicles without having the necessary infrastructure in place to support the increase in demand. Even with the infrastructure they won't be able to keep cost per kWh down as the new equipment will cost a great deal more to permit and construct than the older equipment.

More infrastructure is needed. The cost and timing of the installation of that infrastructure is difficult to ascertain. One thing for certain is that it will be a supply and demand situation. The higher the demand the less supply which will cause the price per kWh to rise and encourage additional supply facilities to be built and the modification of existing facilities none of which will be inexpensive.

California's Pacific Gas & Electric (PG&E) now charges as high as $.42 per kWh, the highest in the state. The rest of the state is at about $0.28 to $0.32.99 per kWh which is no bargain either. Keep in mind that the national average (right now) is about $0.16.54 per kWh. You can expect the national average to approach or exceed the California numbers as new facilities are built or older facilities modified to accommodate EV's.

The cost and timing to build new power generation and distribution equipment will not be inexpensive and will not be readily available. With environmental laws being what they are today just getting the permits to construct new power generation equipment will take years. Procuring the rights of way for new distribution equipment (transmission lines) will also take a great deal of time which assumes it will be available. If a private landowner refuses to provide the right-of-way, his/her land would have to be confiscated under eminent domain which again would take a considerable amount of time. Not to mention trampling of the rights of the landowner. The alternative would be to select another route to by-pass the original Right of Way, again a costly process.

The modification of older distribution equipment will require a lengthy permitting process as well; it will include environmental impact studies which could take years. All of these expenses will be passed on to the consumer. Therefore, you can count on your electric bill to increase considerably over the years, if you own an EV or not.

An article in the "Liberty Bugle Town Hall" identifies the inefficiencies of electric charging stations. As mentioned above, the article says that it could take as much as 13% more power to charge the battery than the capacity of the battery. This is mostly due to the heat generated from the charger which comes in the form of lost kW. An example: a 100kWh battery will take 108 kWh (about 8% more than the capacity) to completely charge due to the heat generated (lost kWh) from the charger in the time it takes to fully charge the battery. They liken it to pouring a couple of gallons of gasoline on the ground when you refuel your internal combustion engine vehicle. It's energy you pay for but get no benefit from." The longer it takes to charge the battery the more heat generated and the more kWh lost.

"So, imagine pouring two gallons of gasoline on the ground every time you fill a 20-gallon tank. People would lose their collective minds. But that will be the standard for every charge of every vehicle in the utopian electric fleet of the future." "The consumer pays immediately for the loss of every kWh that exceeds the actual electricity needed to charge the battery."

"With rates skyrocketing, charging will become unaffordable to all but the most well-off and we've postulated repeatedly that this was always the goal. At the time of the release of the original Green New Deal proposal, whose author (we can't ever let you forget this) said, do you guys think of it as a climate thing? The author continued, "because we really think of it as a way to change the entire economy thing." Biden's "Inflation Reduction Act" is not appropriately named, even Biden admitted that it was mis-named and should have been labeled a "Climate Change Bill". But even that is a misnomer, it should be called the "Economy Destruction Bill".

"No air travel for you and the end of private transportation."

"The goal is to force your mobility profile back to the 19th century. Trains (with wood fired boilers), horses, walking. The only other reasonable explanation is that they are all idiots, and while the rank and file prog parrots might be considered that stupid, the people at the top are not."

The article goes on to say "With electricity prices scheduled to double in New Hampshire (as an example) and the cost of EV's still out of range of most middle and lower-income families, throwing money out the window with every charge might just as well be another tax."

"Line loss or transformer loss is baked into the infrastructure. There is no way to transmit electricity without waste (primarily) in the form of heat. Anywhere from 8-15% or more of the electricity generated by power stations is lost before it gets to you. A carbon footprint we're supposed to ignore."

There appear to be a great many unknowns relative to Electric Vehicles especially in the areas of cost, safety and environmental impact. Imagine trying to evacuate a city or area that has a high category storm approaching like New Orleans when hurricane Katrina hit. Electric cars would compound "Grid Lock" as their charges were depleted. Not the safest place to be! They would essentially turn the highway system into a parking lot.

According to an article in the Western Journal (westernjournal.com) "EV pushers dealt a gut punch; a whopping percentage of EV owners want gas engines back (about 46%). They distrust charging equipment and don't like the range of the EV". In the same article Frank Hanley, senior director of auto benchmarking at J.D. Power, said the EV owners "are experiencing problems that are of a severity level high enough for them to take their vehicle into the dealership at a rate three times higher than that of gas powered vehicle owners". "It is not surprising that the introduction of new technology has challenged manufacturers to maintain vehicle quality" Hanley said. "Another problem is how long does it take to identify the charger operating status? Do you have to wait a half hour to determine if the charger is working"?

While you may save money on scheduled maintenance such as oil changes and other routine periodic maintenance for internal combustion engine vehicles, EV's have an inordinate amount of electronic equipment. If there are problems with that equipment the price tag for repairs is going to be very high. Further, as noted earlier, an EV weighs much more than a gas powered car and will use tires up to three times as fast. A tire rated at 60,000 miles will likely only last 20,000 miles. Tires are not cheap and one of the major components in tires is PETROLEUM. So, if you do away with petroleum, what will your tires be made of and how much will they cost? Not to worry though, China will still have plenty of petroleum, they won't mind selling you your tires.

If you are going to use a public charging system you need to be prepared to spend some time just hangin' around. It takes a lot more time to charge an EV than to fill a gas tank. You might want to schedule a root canal or appendectomy while you wait. Make use of your time. Further, private chargers usually have an app you have to put on your phone.

These companies are now allowing you to charge up to 80% of your car battery capacity. Beyond that they will charge you more per kWh and they will, beyond the 80% mark, reduce the strength of the charger to a "trickle charge" which will likely take longer to "top off" the charge than it did to get to the 80%. The objective is to make room for the next person even if there is no one in line. This will reduce the manufacturer's mileage prediction by 20%. If they say you have a range of 300 miles on a full charge you will have to reduce that number by 20% to a 240 mile range. This of course assumes no air conditioning, headlights or traffic jams.

EV's are very susceptible to hot days or very cold conditions. Brings to mind a story told by an EV owner in Minnesota who had 19 miles left on his car and pulled up to a charging station at a mall; plugged in and went off shopping. It was about 10 degrees F outside, when he returned, about 3 hours later, he looked at his mileage indicator and it said he had 19 miles left on his charge. Very efficient! Either the charger was not operational, though there was nothing to indicate that, or it would not charge in the extreme cold. Also, hot and cold conditions can reduce the range of the vehicle by about 20 to 30%.

One thing to keep in mind is that the mileage rating per charge is based on road trials around a track (usually in fair weather conditions). No traffic congestion, no air conditioning used, no headlights, etc. The true distance available is at best variable. I have had, on occasion, to sit in traffic on LA freeways as much as three hours due to accidents. Many times, you are not 100% stopped, you are inching along as the Highway Patrol is moving 4 or 5 lanes of traffic down to one lane. Therefore, you won't have an opportunity to shut down the vehicle to conserve power. You have to keep moving no matter how slow. How long will the charge last under those conditions?

Another high cost to California taxpayers related to the EV push is the result of having to defend the California Air Resources Board (CARB) against lawsuits over California's push to convert all vehicles to "Battery Operated". A recent article at America Action News (http://american actionnews.com) entitled "Next EV Battle Pits Truckers and Red States Against California-Led Cartel. The article tells a story of how there are lawsuits against the California Air Resources Board (CARB) stating that they have formed "COLLUSIVE partnerships with heavy duty truck manufacturers to eliminate internal combustion vehicles in the state by 2036". They are claiming CARB (run by bureaucrats) has "obtained a cozy cartel arrangement with the manufacturers". I suspect you will see many such law suits emerge over the next several years as the CARB rules negatively impact more industries operating within California. Win or lose these law suits will cost the California tax payer a bundle. They are trying to mandate that any commercial equipment operating within the boarders of California must be electric, even though they cannot supply the power to recharge the equipment. I would guess that California's supply chain and economy will be the first in the nation to collapse. The sad part is that there are so called "LEADERS" in Washington D.C. who think California's plan is a good one for the rest of the nation and are pushing the same agenda.

As mentioned earlier and worth mentioning again, how will all of the mineral excavation, processing, transporting and production for batteries be accomplished without PETROLEUM. All of these processes require diesel fuel.

So, just how economically efficient, safe and environmentally friendly are Electric Vehicles? The answer of course is variable but all things considered I would guess an internal combustion engine vehicle is overall more environmentally friendly and efficient than an EV any day of the week.

So, the question becomes, how did we get to this point where we are subject to overbearing bureaucrats and zealots, who seemingly have no idea of the nation's dependence on petroleum, dictating how we must live our lives in the name of something no one has been able to prove. There has never been any direct link to CO2 and "Climate Change". How exactly is CO2 causing any change in the earth's climate??These same bureaucrats and zealots have no idea of what it takes to make an Electric Vehicle yet they will preach that it is environmentally superior to what we are doing now.

Chapter 3
Atmospheric Composition/ Global Conditions

What makes up the earth's atmosphere? The following is a summary of the gases that makeup the earths' atmosphere which has basically remained unchanged for centuries: Source of the below information: www.geocraft.com "Take the global warming test."

Pay particular attention to the volume of CO2 and Methane!

Imagine, if you will, viewing the atmosphere in "parts per million". Using a vessel capable of holding 1 million equally sized grains of sand; the vessel content would be as follows:

Nitrogen – 76.55 % or 765,500 grains of sand. Color Light blue

Oxygen - 20.54 % or 205,400 grains of sand. Color Pale Green

Argon - .91 % or 9,100 grains of sand. Color Light Tan (sand)

These three gases make up 98% of the earths' atmosphere; the remaining 2% are the so-called "greenhouse gases", without which we would freeze to death. Your vessel is now 98% full.

Next add:

Water Vapor – 1.95% or 19,500 grains of sand. Color white (clouds)

Co_2 - .038% or 380 grains of sand. Color Black ("it's a terrible hazard to humans"- quote source, bureaucrats at the EPA). To define CO_2: it is one carbon atom bound by two oxygen atoms. So, it is two to one oxygen.

Methane - .012% or 120 grains of sand. Color Dark Blue – The chemical symbol for methane is CH_4, 1 carbon atom bound by 4 hydrogen atoms.(the methane number will include miscellaneous gases that are too small to measure here).

Your vessel is now full! If you were to shake the vessel and mix the contents well you would likely not see any of the black or dark blue grains of sand without sifting through the contents.

Now, of the 380 grains of Black sand that represent atmospheric CO_2, 364.8 (96%) are introduced to the atmosphere **"naturally"- volcanoes, sea floor expansion, wetlands and the like**, the remaining 15.2 grains (4%) represent mankind's WORLDWIDE contribution to CO_2 in the atmosphere. This, of course, assumes that mankind is not a "natural" entity in the makeup of life on earth.

As mentioned, the 15.2 grains of black sand represent the complete volume of CO_2 attributed to **GLOBAL** Human Activity. As industrious as the U.S is, according to the Deputy Secretary of the U.S. Department of Energy, under the Biden administration, we, the United States as a whole, are responsible for 13% of the human contribution of atmospheric CO_2. Therefore, we would be responsible for 1.98 grains of Black sand in your vessel.

In order to correct this terrible injustice to the earth the U.S. would have to eliminate all farming, all manufacturing, all food processing, any and all industry. Businesses such as hotels, restaurants, bakeries, yes even "government"; shut down every plane, train, bus, car, motor home, SUV, PRIUS and Tesla. All domestic livestock and pets as well as all wild life would have to be eliminated (every living mammal emits CO2 when exhaling a breath). Homes would have to go unheated and ALL of the 300+ million people in the United States would have to Stop Breathing!!!! The last item renders all conditions above except for animals moot. Once all of this is accomplished you can remove 1.98 grains of Black Sand from your vessel. (Removal of the 1.98 grains assumes you are the last man standing and that you are capable of splitting a .02 sized piece from a grain of sand).

I for one don't believe the American people would go along with what is outlined above, especially the stop breathing part. So, let's look at just one state, California, the so-called "environmental leader", and its contribution to CO2 emissions.

California being a self-proclaimed "environmental leader", once the rest of the country decides against not breathing anymore, let's assume California forges ahead on their own (the Governor claims California will be CARBON NEUTRAL by 2035, a total impossibility). Therefore, with the size of the California population being about 10% of the nation as a whole, once all California industry is eliminated all domestic livestock and wildlife eliminated and all residents Stop Breathing, you can remove .2 grains of Black Sand from your vessel. The .2 grains of black sand would likely be more than offset with Dark Blue grains of sand from all of the methane emitted from the bodies after breathing stops.

A fact that cannot be overlooked here is that CO2 is an absolute necessity for life on earth, without it the earth would be a barren rock. Every breath of oxygen you and all living mammals take will be exhaled as CO2. The CO2 is then absorbed by plant life (CO2 is essential to plant life) trees, bushes, flowers, grain, all agriculture (your food supply). In return the plants, through photosynthesis, produce oxygen for us to breathe. Pretty handy don't you think?

I hope this has put the issue in a usable perspective for you. Whenever possible, introduce "SOUND SCIENCE" to everyone you know. Perhaps we can move forward with a logical approach and eliminate some of the misguided, extremely costly and damaging legislation and "RULES" that are currently being proposed and forced onto the people of this country and the world. The nonsensical legislation for the so-called "Global Warming, Climate Change and/or Climate Emergency" costs U.S. taxpayers more than a trillion dollars annually for an absolute farce.

As mentioned above, one thing to keep in mind as you read on is the protective blanket around the earth that prevents the sun's warmth from escaping is 2% of earth's atmosphere. Removing it would cause us to freeze to death. As outlined above, that 2% is 1.95% water vapor (clouds), the .05% balance being CO2 and methane. The chart below demonstrates the gases that make up earth's atmospheric content. Go to WWW.geocraft.com and click on "Take the Global Warming Test"

Source of Gas percentages: www.geocraft.com

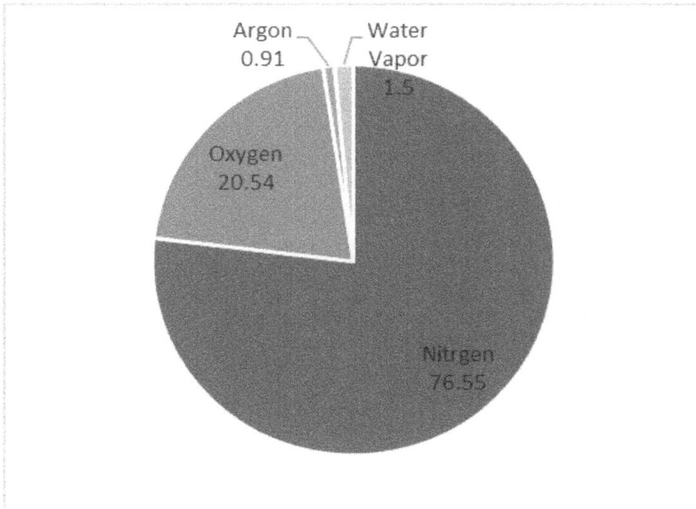

Note in chart above that the CO2 and Methane are so small they don't show up on the chart. They reside between Argon and Water Vapor

If you are interested in what causes the occasional heat wave, look to scientists that study Solar Cycles, not government hacks and ignorant children pushing an agenda.

Methane

Methane is considered another bad actor in the composition of the atmosphere. As stated above, the chemical symbol for Methane is CH4. This translates to 1 carbon atom bound by 4 Hydrogen atoms. Which means that methane is one carbon atom away from pure hydrogen. Methane is NATURAL GAS. The gas many of you use to heat your home or cook on a gas stove or heat your bath water is mostly methane. The residential and industrial gas stream is made up of about 90% to 93% methane. Various percentages of butane, propane, hexane and trace amounts of CO2 and nitrogen make up the balance of the gas delivered to the customers of gas companies. The additional gases in the residential stream are "BTU enhancers" (British Thermal Units) as they burn hotter than methane they increase the temperature at the burner tip. Methane is relatively cool burning gas at 1000 BTU per cubic foot. Whereas, Propane burns at more than two and a half times the temperature of methane, slightly more than 2500 BTU per cubic foot.

As mentioned above, the methane contained in the atmosphere is about .012% or 120 grains of sand in your vessel.

The Santa Barbara Channel off the coast of California is said to be the largest methane seep in the world. I have not compared the emissions against other seeps but I know there is a large amount of gas seeping into the atmosphere from the seafloor. You can physically see the bubbles of gas come to the surface when fishing or boating in the Channel.

To demonstrate what happens in the channel there was a project in the early 1980's sponsored by Atlantic Richfield Company (ARCO) to capture some of the seepage. The motivation for ARCO was the opportunity to expand their Carson (LA) refinery capacity using offset emissions credits. An agreement was made between ARCO and the

California government environmental entities. Arco would gain emissions credits equal to 50% of the oil and gas captured over a given time frame. They constructed two pipe "Pyramids" 50' high by 100' wide at the base. The pyramids were lowered to the seafloor to capture oil as well as gas. The oil and gas were then transported to a platform facility in the channel by pipeline. To date the pyramids have captured 7.6 **BILLION** cubic ft. of gas, the amount of crude oil captured is unknown but it is estimated at more than 25 barrels per day. The oil recovered would be in excess of a half - million barrels since the activation of the pyramids 44 years ago.

The Santa Barbara Channel seeps 3 to 7 million cubic feet of natural gas (Methane) per day and 100 to 175 barrels of crude oil per day, about 55,000 barrels of oil per year. To keep up with typical media reports that equates to 2,310,000 gallons of oil per year which stays in the ocean. The methane gas rises to the atmosphere. To put things in perspective the State of California just settled a lawsuit, for $73 million, against a pipeline company for spilling 126,000 gallons, 3,000 barrels (about 0.05% of annual seepage) of oil off the coast of Santa Barbara.

The natural seepage is due to reservoir pressure. The Santa Barbara channel is home for one of the largest oil reservoirs in the country and it is a larger gas reservoir than it is oil containing trillions of cubic feet of methane. The pressures of the undersea oil and gas reservoirs force the oil and gas out through fissures in the sea floor. Drilling into a high-pressure reservoir and producing the hydrocarbons would reduce the pressure thereby reducing the seepage into the atmosphere and bring useful fuel to onshore communities. Yet the state of California and the

Federal government have banned drilling off the coast. There are oil platforms in the channel that were put in operation decades ago. Back in the 1990's the city and county of Santa Barbara were up in arms about getting the oil industry out of their area. One Santa Barbara resident was asked how he thought of the oil company's performance relative to operating the platforms. His response was "I think they have done a great job, I just don't want them here". (NOT IN MY BACKYARD)

There are additional seeps around the world. There is another large seep in Alaska in the North Slope region, also banned from drilling. If Methane is such a problem in the atmosphere why does the government refuse to allow drilling to reduce the natural emissions and capture the gas for practical use. Emissions from the gas in home appliances and industrial applications do not measure near as much environmental impact as does the raw material coming from the seeps.

What of man made seeps also known as "Land Fills"? There are landfills across the country that emit methane gas due to decomposition. Some municipalities have launched efforts to capture the emissions and use the gas for fuel. This is not something that can be done on every landfill but it will remain unknown unless someone investigates. There are issues associated with such a project as "reservoir" pressure may not be consistent and flow may be sporadic. However, it is a source of energy that could be collected and stored at a facility that would provide additional equipment to enhance delivery possibilities. Have to think out of the box!

This leads me to believe that methane in the atmosphere is not a problem (about 120 grains of sand in the vessel). It is just something for doomsday politicians to use to frighten the unknowing general public and for EPA bureaucrats to use to destroy the Oil and Gas Industry by placing onerous controls on the industry costing billions of dollars per year which translates to higher energy cost for the consumer.

It has gotten so ridiculous that now the State of Washington is trying to pass legislation to tax farmers for cows flatulating. How in the world could anyone possibly measure that??? Are they going to have to design and place a metering system on every cow's butt?? Talk about dealing in minutia. Mental illness is alive and well in the Washington State legislature. These people are idiots!

The atmosphere consists of the gases mentioned above and other than minute changes it has remained consistent for centuries. Where are these so-called scientists finding excessive CO2 levels, >.4% is all there is. If they force additional volume in a computer model then their results are skewed and cannot be viewed as factual by any stretch of the imagination.

What Causes Changing Weather Conditions and False Proclamations

There are a number of factors that impact the weather. There is polar shift, Solar Cycles, sun spots, the earth's "wobble", gravitational pull from the moon and sun, jet stream shift, to name a few, all natural conditions and nothing mankind is causing nor can mankind prevent. A recent article in "BusinessToday.in" speaks of polar shift saying that "the earth's magnetic poles are shifting and could swap places". They claim that "this has happened 183 times in the last 83 million years". No one has identified the timeline for the swap. Don't know what the impact on the weather would be but one thing for sure it would not be caused by human activity nor will it be prevented by mankind. Politicians will blame it on human activities that they, the politicians, must somehow stop!

If you listen to politicians and "master scientific celebrities" such as Hanoi Jane Fonda and Ted Danson the world is going to end very soon (according to their predictions it already has) but they have nothing to support their claims other than "Scientists Say". Ted Danson's, 1988 prediction (with no factual support for the claim) that in ten years from that time the oceans of the world would be dead and all sea life with It. So far, we are 26 years beyond that deadline and it has yet to happen. There were predictions that Glacier National Park would be void of glaciers by the year 2020. The Forest Service even put up signs telling visitors to take pictures now because the glaciers will be gone by 2020. Glaciers still going strong at this time (more on these issues in the next chapter).

John Kerry and Al Gore, possibly the largest frauds in the scheme, fly around the world to one conference after another preaching the same misinformation (LIES) as the rest of the woke politicians and celebrities. They start with "Scientists Say", then branch out with one absurd observation after another. They never identify the scientists just what "THEY SAY". If in fact there are scientists that support the politicians position you can bet that they are either government employees or they are dependent on government grants for their livelihood. If they disagree with the political mantra and tell the truth, they will be fired or their grant funds will be cut off.

I find it humorous that when Al Gore would make his trips around the world in the winter to speak on global warming he would be greeted by a blizzard. This would happen whether he was in the U.S or Europe. He would leave his limo parked outside of the venue running with the heater on so he would be warm when traveling back to the airport to a private jet. This shows true dedication to reducing emissions.

Back in the 1970's we were going to experience the next ice age, starting in about 30 years. Well, that ice age then morphed into global warming in the 1980's. Global Warming was introduced just two years after the worst blizzard in U.S. history struck New England states (February,1978).Then it was that global warming was going to cause the next ice age. When in the 1990's the warming turned out to be cooling. By the early 2000's the mantra then became "CLIMATE CHANGE". You can't deny it, the climate is changing. Well, here is a news flash, the climate changes every single day no matter where you are. There is nothing static nor repetitious about climate cycles, other than typical Four Seasons activity. The temperature goes up and it goes down and it will not be the same on the same day of any given year regardless of the season. There is nothing the government is going to do about it regardless of how powerful elected officials think they are.

In a recent video by Wide Awake Media, Geologist Professor Ian Plimer, a REAL scientist, "blows the human induced "Climate Change" myth completely out of the water. He explains "that we are actually in a very boring time in terms of period extinction. We are in an interglacial period which produces a warmer climate". Enjoy it while it lasts, in a few thousand years there will be another ice age, keep your parka packed and ready. (Wikipedia states: "An interglacial period is a geological interval of warmer global average temperature lasting thousands of years that separates consecutive glacial periods within an ice age").

The professor goes on to explain "Changes in the climate are driven by the great ball of heat in the sky, which we call the Sun. A TRACE GAS CO2 HAS NOTHING TO DO WITH OUR CLIMATE. If we get further away from the Sun, it gets cooler and if we get closer to the Sun it gets warmer. No legislation from the UN (or U.S.) can change the orbit of the earth".

Al Gore predicted the ice caps at the north and southpole would be melted by now as did John Kerry. Gore also declared that the "DEBATE IS OVER" no one can say when it started, just a tactic to shut down the opposition. As mentioned above as a result of the declaration relative to the ice caps, sea levels would rise high enough to put the Maldives and Samoan Islands underwater (Water World anyone?). That has not happened. Recent studies show Samoan's are not living on houseboats and rafts and there are new resort projects being built in the Maldives. These climate "experts" (I have more colorful and descriptive names for them) talk of rising sea levels and blame it on ice melt. Yet NASA satellites show the Arctic ice cap has expanded by 23% since Mr. Gore's prediction.

I would propose an alternate cause of rising sea levels, that would be sea floor expansion. We have all seen the Hawaiian volcanoes spew molten lava into the ocean as it increases the size of the island. In addition, there are undersea volcanoes that do the same thing. As we know, volcanic activity is actually the way the Hawaiian Islands as well as many other Islands were formed. Don't know what the sea levels were at the time but I would venture to guess that the sea levels rose to some degree to accommodate the land mass. There is an undersea volcano 25 miles south of the big island of Hawaii that is creating a new island that is right now less than a mile below the surface. The volcano Kamaehuakanaloa, (no I can't pronounce it either), is creating the next Hawaiian Island, albeit about 10,000 to 100,000 years from surfacing. It lends itself to the theory of seafloor expansion and sea level rise.

There is another undersea volcano called the "Axiel Seamount" it is located about 300 miles off the coast of Oregon. According to William Chadwick, a research associate at the Oregon State University in an interview with ABC News, "the volcano is displaying behavior that indicates an eruption is eminent in 2025". You can find the story here: https://abcnews.go.com/US/mile-long-underwater-volcano-erupt-off-west-coast/story? This volcano has erupted several times in the last 30 years each eruption spews molten lava into the ocean which then becomes a permanent part of the seafloor. So, you have one underwater volcano in the Hawaiian Islands and one off the coast of Oregon. I have done a little research on the subject and found that there are others scattered about the Pacific and other oceans of the world. There is now a new Island that has surfaced off the coast of Japan. The island was created by an undersea volcano that has now surfaced and will continue to grow, displacing more sea water. You would think that this activity would lend itself to the "seafloor expansion theory" for rising sea levels.

Sea levels rise constantly, by minute amounts, by way of seafloor expansion. Aside from the large volcanic eruptions, there is constant activity of molten lava being injected into the seafloor through fissures, the lava is then immediately cooled by the ocean at which time it solidifies as part of the sea floor. As noted above, this doesn't just occur around Hawaii or the Axiel Seamount off the coast of Oregon or the new island off the coast of Japan. The entire Pacific Rim as well as other oceans of the world have molten lava being injected into the sea. Tectonic plates are another example of seafloor expansion.

To confirm, see the excerpt from Wikipedia below:

"A list of active and extinct

submarine volcanoes and **seamounts** located under the world's oceans. There are estimated to be 40,000 to 55,000 seamounts in the global oceans.[1] Almost all are not well-mapped and many may not have been identified at all. Most are unnamed and unexplored. This list is therefore confined to seamounts that are notable enough to have been named and/or explored". Google "Wikipedia submarine volcanoes" for a list of undersea volcanoes / seamounts. The list is long!

There is never any mention of "SUBMARINE VOLCANOES OR SEAMOUNTS" when you hear talk of rising sea levels! Do you think these volcanoes could have something to do with a change in sea levels? It would be like dropping ice cubes into a glass of water as the level in the glass rises. So, I would argue that rising sea levels are a product of seafloor expansion, not my SUV. Studies show the ocean has gained about 3.5" in the last 30 years. Not quite enough to submerge the Samoan Islands. If you calculate the per year exposure it would be about .12 inches per year (do you think 55,000 undersea volcanoes could be responsible?) If this is consistent you could expect sea levels to rise an additional 3 inches by 2055. Again, not quite enough to submerge the Soloman Islands. I am told that these sea level measurements are accomplished by satellite.

What about ships and boat hull displacement? All ships and boats displace water, again like putting ice cubes in a glass of water. The water rises when you introduce a ship or boat, albeit a small amount but when you consider all of the ships and boats in ports and marinas around the world it is more significant. I would venture to guess it could cause a measurable rise in sea level. A fully laden cargo ship would displace a much larger volume of water than an empty ship. Are these measurements included in the sea level calculations? Doubtful!

Another misinformation episode, in 2004 Southern California had an unusually warm July. There was a newscast in Los Angeles where the news anchor blasted out a statement that "this was the Hottest July in 300 years". I seriously doubt that anyone was measuring and recording the temperature in Los Angeles 300 years ago. If so, who was recording it? What did they use to measure the temperature? How accurate was the equipment used? Where did they record it? Where is the record kept? The media just spew out unverifiable statements and expect the public to absorb it. Sadly, there are those in their audience that are that gullible. The absurd statements generally go unchallenged.

There are many ways to measure temperature, the most common for many years throughout history was accomplished using mercury thermometers while fairly accurate they were difficult to read. The mercury thermometer was invented in the year 1714 by Daniel Gabreil Fahrenheit, a German physicist. In 1724 he invented the Fahrenheit temperature scale. Note that 300 years prior to July 2004 (1704) the mercury thermometer was not yet invented nor was the Fahrenheit temperature scale. It is difficult to understand where our astute LA news anchor sourced his information and could possibly make the claim that July 2004 was "the hottest in 300 years".

Something to consider when hearing of higher temperatures in cities is the time frame being considered in the measurement. New construction in most large cities in the country consists of concrete and mirrored glass exteriors. The older buildings would have been made of brick, stucco or wood and they had clear glass windows the construction materials would in fact absorb heat and cities were much smaller. Mirrored glass reflects heat. Further, the cities, for the most part, are much larger than they were years ago. An example of that would be Los Angeles which has had enormous growth in the last 75 years. An airline pilot once told me that when landing at Tokyo International Airport you see a sign saying "Welcome to Tokyo"; the other side of the sign says "LA City

Limits". A bit of an exaggeration but you get the idea. Therefore, depending on where the measurement is taken the temperature will read higher than true in a modern city compared to just a few miles away in the suburbs. Research tells me this phenomenon is called UHI, "Urban Heat Island".

The environmental zealots claimed Hurricane Katrina was caused by carbon emissions from internal combustion engines which caused the "Climate Change" therefore the hurricane. Can anyone tell me how much CO2 was contained in the hurricane??? Forest fires are also blamed on carbon emissions. It is difficult to understand how plant life that is absolutely dependent on CO2 could create a fire due to its presence.

I will not dispute that there is evidence of global warming. Professor Ian Plimer article mentioned above would confirm that. I for one am thankful for the warming. There is, however, evidence that Venus, Mars and other planets have increased slightly in temperature as well. I'm pretty sure my SUV didn't cause it. If you look at the earth's temperature 18,000 years ago you would find it much colder and the majority of the North American continent would have an excess of 1,000 feet of ice covering the surface that was one huge glacier. Incidentally, the CO2 content in the atmosphere during that ice age was almost 15 times higher than it is today. Why? There were no trees and other plants to absorb the earth's natural emissions of CO2. If you are concerned about CO2 emissions, plant a tree! The earth has also experienced short term ice ages since the last long term one. George Washington was dealing with a mini ice age while fighting the revolutionary war. Old George didn't have an SUV and I doubt that his horse put out that much gas! It has since warmed up. These swings in weather are not man made, they are natural occurrences.

What you see propagated by uninformed politicians and so-called "climate experts" are the result of computer models whose output is entirely dependent on the input. What typically happens is they feed the program with a myriad of "WHAT IF'S" meaning they are suppositions and some outright impossibilities yet they will portray the results as factual information. It's rather like the creative accountant that says "what do you want the numbers to look like"??

A new scare introduced by "The Guardian" which claims to be the news website of the year. Says "by 2050, millions of people will be displaced because of rising sea levels due to global warming". (I would add that millions of people have been displaced due to Biden's open borders policy). This brilliant revelation comes to us by the "Union of Concerned Scientists" (USC). This study says that "1,100 Coastal communities, primarily in Florida and California with the inclusion of some other coastal states like Massachusetts and New Jersey. All will see flooding on a monthly basis and that 934 of the communities will see flooding every other week" Huh? Will the ocean level change every other week? All of these doomsday predictions seem to have the same common denominator, 2050 the year of the END. By 2050, a date forecast far enough out in the future that once it arrives, with minute changes in sea level, you will have forgotten the predictions of 2024.

Once again, the "study" above was likely conducted using computer models where supposition (what if's) guided the input therefore the output would be speculative. The consensus for most if not all of these studies appear to show 2050 as an extremely catastrophic year or many catastrophic years leading up to that date. 2050 replaced 2020 as the last doomsday date; 2020 replaced 2016 which replaced 2012 which replaced 2008 and so on, all the way back to the 1950's. Since 2000 and 2020 passed without incident therefore the predictions fizzled out. The worlds' younger generations are not familiar with the prior predictions, only those for 2050 so we really need to look at 2050 (according to predictions).

The so-called "greenhouse gas" blanket around the earth, as mentioned above, is made up primarily of water vapor and is only 2% of the atmosphere that has not changed much since the last ice age. Natural causes are the majority of the makeup of CO2 (96%). Mankind is not destroying the earth by producing CO2 and mankind cannot change the natural order of weather events.

As young boy, growing up in New Jersey, our family would vacation at Seaside Heights on the Jersey Shore. I can remember on some occasions where we experienced water levels rising and we waded in the streets inland from the beach (a few inches of water), the beach and some of the boardwalk being underwater for a brief time. The water levels rose because of storm surges and/or plus tides. Nothing to do with global temperature.

Environmental activists claim that we are seeing more intense storms with more frequency than the past. Not so! If you look back more than five years (most people won't, they would rather believe politicians and activists who are either uninformed or just lying) and study what has happened weather-wise over the last 100+ years or so, you will see that we have had several category five hurricanes that made landfall in the U.S. Also, you would see that there have been massive tornadoes throughout history. Case in point on September 8,1900 a category 5 hurricane made landfall in Galveston, Texas. That storm killed thousands of people. The modern day storm warning systems were not yet developed therefore residents did not know to evacuate.

Fortunately, we have systems in place that forewarn the population of areas where hurricanes and other storms will strike. The systems are in place to save lives and they work. The difference you see now is in the amount of storm damage done today as opposed to the past. The reason is the population in the storm's path is considerably higher than it was historically due to Real Estate developments in areas with potential to flood in heavy storms or where a tornado may find a path. A tornado path is always unpredictable which is why historically, people building houses in tornado prone areas included storm cellars in their construction plans (remember "The Wizard of OZ"). Large developers don't spend the money to do that. The cellars would not reduce surface damage but certainly save lives. The above is food for thought. The human population is not causing "Global Warming". You will hear all manner of doomsday predictions from excessive temperature to extreme sea levels. An example is the "brilliant scientist" that recently predicted that the sea level on the Atlantic coast would rise to the point of the left elbow of the Statue of Liberty by 2050, which equates to 241' or 9.64' of sea level rise every year for the next 25 years. More on that in the next chapter.

Try applying a little common sense and some valid research when hearing these absurd claims and predictions. Don't just take it for granted that these people who claim to be "scientists" must be right. There are many motivators for people to sign on to these predictions most of which are monetary and have nothing to do with the truth.

Chapter 4
Scientist Say

Not all who "say" are scientists. Through the years we have heard a number of dire predictions relative to the devastation that will occur due to human activities associated with "the next ice age, global warming/ climate change". The predictions are generally prefaced with "Scientists Say" or "They Say" however, scientists or their organizational affiliation are rarely mentioned by name. Not only have we heard about global warming but also global cooling, predictions from the early 1970"s, the next ice age will be upon us by the late 1990'S (the east coast blizzard of February 1978 would support that theory), the earth's oceans are dying at an alarming rate and will be dead by the year 2000, glaciers disappearing by 2020, polar ice caps melting and will be gone by the turn of the century (advent of the 21st century) and so on. Today you can find predictions that the world's oceans are going to rise very quickly and by 2050 there will be no beaches on the Jersey Shore and Manhattan Island will be underwater, these predictions come from computer models which are generated by "What if" data input. This discounts the fact that the oceans have gained about 3.5"'s in the last 35 years, which equates to .1" per year. (Sea floor expansion and volcanic activity?).

There were signs placed at Glacier National Park that said "take your pictures now because the glaciers will all be gone by the year 2020". The signs have since been removed as the glaciers are still there.

Many predictions from brilliant scientific Hollywood celebrities making claims as in the 1988 prediction that the oceans of the world would die by the year 2000. We know that did not happen. Oceans and sea life are still going strong.

Enter Al Gore the ever-vigilant environmentalist that lives in a 20,000 Square foot mansion with a carbon footprint that exceeds that of 99% of the U.S. population. In January 2006 Gore predicted that we had "ten years to save the planet". In addition, the Samoan Islands would be under water at around the same time. Meanwhile he is forced to fly on private jets in order to educate the world on global warming or now climate change. He produced a film that was dubbed "An Inconvenient Truth" which should have been named "Very Convenient Lies". In the film he shows an emaciated Polar bear lumbering across the ice pack on the verge of collapse, claiming that arctic temperatures are driving the polar bear population to extinction. The bear in the film may very well have been a very old or very sick bear, it had nothing to do with arctic temperatures. As with all mammals they have a lifespan. Photographing a dying animal and blaming it on the climate in an attempt to further an agenda is not only irresponsible but also immoral. Research shows that the current population of polar bears is thriving, the "International Union for the Conservation of Nature shows a mid-point estimate at 26, 500 (range: 22,000 and 31,000) bears in 2015". Considerably more than the population of the 1960"s which is said to be below 10,000 bears.

As Al Gore jets around the world preaching doomsday based on global warming one might want to look into how he started on that mission. When Gore lost his bid for the presidency of the United States, he was looking for another spotlight and road to riches. He with a partner in a company called Generation Investment Management (GIM) hung their hat on the U.S. congress passing Cap-and-Trade legislation where GIM would, working through the Chicago Climate Exchange program, trading carbon credits for profit. The above is a very simplistic explanation for the process but the actual structure for trading carbon credits can get complicated. Regardless, the effort failed and Gore decided to go after the Global Warming "Golden Goose". Starting with speaking arrangements where he would preach pending catastrophic doomsday events. Which then morphed into the movie mentioned above. I might note that none of his predictions have come to pass. And Hollywood has egg on their collective faces for giving Gore an Academy Award for Documentary that was a purely fictional movie.

The coming ice age was forgotten and soon morphed into global warming and that we would see an increase in global temperatures by 2 degrees Celsius by the end of the 20th century. That time frame has now been moved up; it is now predicted that that will happen by 2050 a good 50 years in the future from the last drop dead date of 2000. They continue to move the goal post. We are now half way there with no significant changes.

Enter Greta Thunberg, the 14 year old "SCIENTIST" and Swedish grinch. She dropped out of school at age 11 and would not speak to anyone but her family for three years, she was diagnosed as having Aspergers Syndrome, a form of autism spectrum disorder. Then at age 14 she emerged as a "climate expert" and media darling. No idea where her education came from. A quote Greta made in June of 2018, once again no scientist named, "A top scientist is warning that climate change will wipe out all of humanity unless we stop using fossil fuels over the next five years". It has now been over 6 years since that declaration with no change in humanity except for more brain-dead people taking up the cause.

Greta decided that she would not fly on commercial airlines because of the carbon footprint when traveling to the U.S for a climate related function, instead she boarded a sail boat to cross the Atlantic. The crossing was highly publicized. There was absolutely no consideration that the hull of the boat was likely fiberglass which uses epoxy resins made with petroleum products, all of the wood that might have been aboard was treated using hydrocarbons, all of the metal used for cleats and all of the necessary sailing hardware were mined, processed and shipped using hydrocarbons. In addition, the sails were likely nylon based and the vessel probably had a gasoline or diesel-powered engine to navigate in marinas. All the clothes she and the crew wore would also be petroleum based, nylon, rayon, dacron, etc. What it boils down to is these self-declared experts are nothing but frauds looking for notoriety and to profit from a public that is not willing to fact check their claims.

We have seen and heard predictions of the end of the world as we know it since the 1950's. Actually it dates back to the early 1900's but those predictions have little to do with what is happening today. Many of the predictions are quoted below in chronological order. It was said in 1958 that the North Pole ice cap would melt. In 1968 it was too late to save the earth, humans have destroyed it beyond recovery. In 1970 the new ice age would be upon us before the turn of the century. In 1971 we would be experiencing a new ice age within 50 to 60 years. In 1978 the new ice age was coming fast. In 1979 the North Pole would be melted (what happened to the ice age?). In the 1980's the mantra suddenly changed from an ice age to the new mantra of global warming, melting polar ice caps causing dramatically rising sea levels. In 1988 the Maldives would be under water as well as Al Gore's prediction that the Samoan Islands would be under water. In 1988 as noted above Ted Danson said the oceans would be dead in ten years with humanity to soon follow. All of these predictions are false flags that were produced by computer models filled with bogus data in order to come to the desired conclusion for political purposes.

The Global warming mantra of the 1980's suddenly changed to "Climate Change" when the temperatures in the late 1990s and early 2000s cooled by a few degrees overall. So, the only saving grace was to declare "Climate Change" that has now morphed into a "Climate Emergency". Some politicians and bureaucrats will employ any and all fear tactics to get the public to see things their way regardless of facts. If you quote facts, you are in "denial". All that is necessary for them to be right is "SCIENTIST SAY". On very few occasions you may hear them identify a scientist by name. You can bet, however, that the scientists that "say" are dependent on the government either from direct employment or via government grants for their livelihood.

An article at "https://Americanbriefing.com Bullying-harassing-name-calling-epa-retaliated-against-three-dissenting-scientists-inspector general-finds/" "There was a recent incident in Washington D.C. where three government employed scientists dare to counter the general thinking of the EPA relative to issues regarding policies around chemicals. All of them were faced with bullying, harassment, name calling, etc". Can you imagine what would happen if they came out against CLIMATE CHANGE.

John Kerry, another brilliant scientist said in 2009 "the arctic would be ice free by 2013" Still plenty of ice in the arctic!

The so-called science these people are using is a miniscule snap-shot in time and using computer models to enhance and support their predictions. Computer models are only as good as the information supplied, "garbage in, garbage out". We didn't even have weather satellites, until the 1960's. Those early satellites were not capable of measuring temps around the world, the 1970's brought more sophisticated equipment. Modern day satellites do not measure global temperatures, rather they measure atmospheric brightness. Scientists then use computer models to translate the brightness to temperature (how that works I don't know). Prior to satellites scientists depended on core sampling, tree growth rings and other forensic methods to identify historic temperatures and atmospheric C02. Incidentally, at the risk of being repetitive, the atmospheric CO2 content during the last ice age, 18,000 years ago, was almost 15 times higher than it is today. Why? No trees or other plant life to absorb natural production of CO2.

The earth has been warming, this last cycle, for approximately 18,000 years, since the last major ice age and my SUV didn't cause it. As mentioned above If it hadn't warmed the majority of the North American continent would still be under 1,000 feet of ice. The most accurate data relative to current temperatures come from satellite technology and that has only been around for about 50 years a mere blink of an eye in relation to the earth's history or even the last ice age.

As mentioned, in a previous chapter, while watching the news in Los Angeles in early August 2004, the news anchor stated that Los Angeles had the hottest July in 300 years. That ludicrous statement went completely unchallenged. Where the media is concerned, they will take the side of politicians and so-called environmentalists every time. Many times, they make up things as they go. They will never do the research to prove the "Global Warming / Climate Change" claims true or false. It is much easier and more popular to go along with the political narrative regardless of facts (helps ratings). Keep in mind that none repeat NONE of the catastrophic events predicted for the 20th century and early 21st century said to occur by 2020 have materialized. The fact is that NONE of the decades old catastrophic predictions have even come close.

A declaration was recently presented to the world by 1,960 real Climate Scientists that says "There is no Climate Emergency". The group is called CLINTEL and they claim that the climate emergency is a myth. Source: CLINTEL.org.

Dr. John F. Clauser, winner of the 2022 Nobel Prize in physics, and an ambassador of CLINTEL's World Climate Declaration, in addition Dr. Clauser is on the Board of Directors of the CO2 Coalition (https://co2coalition.org/facts/ :

"The popular narrative about climate change reflects a dangerous corruption of science that threatens the world's economy and the well-being of billions of people. Misguided climate science has metastasized into massive shock-journalistic pseudoscience. In turn, the pseudoscience has become a scapegoat for a wide variety of other unrelated ills. It has been promoted and extended by similarly misguided business marketing agents, politicians, journalists, government agencies, and environmentalists. In my opinion, there is no real climate crisis. There is, however, a very real problem with providing a decent standard of living to the world's large population and an associated energy crisis. The latter is being unnecessarily exacerbated by what, in my opinion, is incorrect climate science."

Factual information regarding the climate can be found on the CLINTEL website noted above. There is a wealth of information on the website for CLINTEL. I would encourage anyone with an interest in true science to visit the site.

As noted above the CLINTEL organization published an article entitled "The World Climate Declaration". The main point of the article is that there is NO climate emergency! Below are a few excerpts from the Declaration:

"Natural as well as anthropogenic factors cause warming".

"The geological archive reveals that Earth's climate has varied as long as the planet has existed, with natural cold and warm phases. The little ice age ended as recently as 1850. Therefore, it is no surprise that we now are experiencing a period of warming".

"Warming is far slower than predicted"

"The world is warming significantly less than predicted by IPCC (Intergovernmental Panel on Climate Change) on the basis of modeled anthropogenic forcing. The Climate models have many shortcomings and are not remotely plausible as global policy tools. They blow up the effect of greenhouse gases such as CO2. In addition, they ignore the fact that enriching the atmosphere with CO2 is beneficial".

"CO2 is plant food, the basis of all life on earth".

"CO2 Is not a pollutant. It is essential to all life on Earth. Photosynthesis is a blessing. More CO2 Is beneficial for nature, greening the Earth: additional CO2 in the air has promoted growth in global plant biomass. It is also good for agriculture, increasing the yields of crops worldwide".

"Global warming has not increased natural disasters"

"There is no statistical evidence that global warming is intensifying Hurricanes, floods, droughts, and suchlike natural disasters, or making them more frequent. **However, there is ample evidence that CO2-midigation measures are as damaging as they are costly".**

The EPA is trying to force CO2 sequestration across the nation under the guise of protecting the planet. This is a very expensive proposition and will cost consumers and taxpayers billions of dollars.

“Climate policy must respect scientific and economic realities”

“There is no climate emergency. Therefore, there is no cause for panic and alarm. We strongly oppose the harmful and unrealistic net-zero CO2 policy proposed for 2050. If better approaches emerge, and they certainly will, we have ample time to reflect and re-adapt. The aim of global policy should be prosperity for all by providing reliable and affordable energy at all times. In a prosperous society men and women are well educated, birth rates are low and people care about their environment”.

Bottom line is CO2 Is not only necessary for life on earth but beneficial to plant life which is beneficial to human life. CO2 Is not causing weather conditions to increase in intensity or frequency.

A brief Summary of Historic Predictions

(Some of the below are repeated from above listings but with more detail)

” Below are a number of predictions for the planet from the 20th century as well as some for the 21st century, note some of the prominent sources of the information. Organizations such as NASA, National Geographic and highly respected Universities:”

Source: Roman Balmakov,

You Tube – "Inconvenient Truth 32 Climate Predictions Proven False/ Facts Matter https://www.youtube.com/watch?v=E1e5HAZo4iw"

"In 1958 an article in The New York Times stated Some scientists estimate that the North Pole pack ice is 40% thinner and 12% less area than it was a half century ago. And that even within the lifetime of our children the arctic ocean may open, enabling ships to sail over the North Pole".

Note: "at the time the Arctic ice pack was 7' thick, as of March 2023 the Ice pack is 7' thick".

1967 – The Salt Lake City Tribune reported that it is already too late for the world to prevent a long period of famine according to research biologist Paul Ehlrich "Population Control or the Race to Oblivion". "The Population Bomb". "Along with the prediction he recommended sterilization drugs be introduced to Americans food and water supplies."

1970 – April – The Boston Globe reported – "Scientists predict a new Ice Age by the 21st century citing pollution EXPERT James Lodge. They continued, "air pollution may obliterate the sun and cause a new Ice Age in the first third of the new century". Well, we are three quarters into the first third of the 21st century with no sign of an Ice Age.

1971 – The Washington Post – "U.S scientist sees New Ice Age coming. The world could be as little as 50 or 60 years away from a disastrous new Ice Age, a leading atmospheric scientist predicts". (No one named just a leading atmospheric scientist)

1972 – Brown University – "A conference attended by 42 TOP America and European "investigators" concluded a global deterioration of the climate by orders of magnitude larger than any hitherto experienced by civilized mankind is a very real possibility and indeed may be due very soon. The present rate of cooling seems fast enough to bring glacial temperatures in about a century if continued at the present rate". What exactly are American and European INVESTIGATORS???

1974 January - The Guardian newspaper – "Space satellites show new Ice Age coming fast!"

1974 June – Time Magazine – "Another Ice Age?" "Tell-tale signs are everywhere from the unexpected persistence and thickness of pack ice in the waters around Iceland to the southern migration of a warmth-loving animal like the Armadillo from the Midwest". Okay!??? Was this a saving the Armadillo effort?

1978 – New York Times – "An international team of specialists has concluded from eight indexes of climate that there is no end in sight to the cooling trend of the last 30 years. At least in the Northern Hemisphere". February of 1978 was the year of the worst blizzard to hit New England in the history of the U.S., still the worst on record.

A SUDDEN SHIFT

1979 February – New York Times- "There is a real possibility that some people in their infancy will live to a time when Ice at the North Pole will have melted. A change that would cause swift and perhaps catastrophic changes in the climate”. What happened to the Ice Age? I missed it, guess I’ll have to pack up my ice skates!

1982 May – New York Times – “Citing the director of the United Nations Environmental Program, if the world doesn’t change course, it would face an environmental catastrophe which will witness devastation as complete, as irreversible, as any nuclear holocaust by the year 2000”. Just how, exactly, does the “world change course”?

1988 September – “There was a report that said the Island nation of the Maldives is at risk of becoming completely covered by a gradual rise in average sea level in 30 years and the Maldives and its people could come sooner if drinking water supplies dry up in 1992 as predicted”. Not sure what all that means.

1989 – San Jose Mercury News – “A senior environmental official at the United Nations, Noel Brown, says entire nations could be wiped off the face of the earth by rising sea levels if global warming is not reversed by the year 2000”. I guess the result of another “SCIENTIFIC” computer model.

2000 March – The Independent – “Snowfalls are now just a thing of the past; children just aren’t going to know what snow is within a few years. Winter snowfall would become a rare and exciting event”. No idea what this guy was smoking.

2001 – Albuquerque Journal – “The change in climate could potentially extirpate the sugar maple industry in New England within 20 years according to George Hurtt, co-author of a 2001 Global Warming Report commissioned by the U.S Congress". I guess they eat a lot of pancakes in Albuquerque to be concerned. Maple syrup industry is still going strong in New England! Congress making the best use of your tax dollars!

2004 February – "A secret Pentagon Report that predicted climate change will lead to nuclear war; major European cities will sink into the ocean and Britain would descend into Siberian climate by the year 2020". That would likely freeze BIG BEN. Where did this report come from??? More of you tax dollars at work.

2006 January – Al Gore – "Unless drastic measures to reduce greenhouse gases are taken within the next 10 years the world will reach the point of no return". Return to what I don't know!

2007 – The New York Times – quoted the head of the United Nations Climate Panel ""this year was the defining moment of the Climate Change fight, if there is no action before 2012, that's too late". Too late for what?

2007 – An article in CanWest News Service – "The Arctic Ocean could be free of ice in the summer as soon as 2010 or 2015, something that hasn't happened in more than 1 million years". Did my SUV cause that a million years ago??

2007 December – Associated Press – "NASA finding – Arctic Sea ice gone in summer within 5 years- at this rate the Arctic Ocean could be nearly ice free at the end of summer 2012".

2007 December – A scientific Reporter for the BBC San Francisco – "Artic summer ice free by 2013. Our projection of 2013 for the removal of ice in summer is not accounting for the last minima in 2005 and 2007- so given that fact you can argue that maybe our projection of 2013 is already too conservative".

2008 March – A Norway Publication, name unknown – "If Norway's average temperature this year equals that of 2007 the ice cap in the Arctic will all melt away, which is highly possible judging from current conditions" The temperature in 2008 was slightly higher than 2007 yet there is still an Arctic ice cap.

2008 April – New Scientist Magazine – "The North Pole could be ice free in 2008 there is this thin first year ice even at the North Pole at the moment that raises the spectre, the possibility that you could become ice free at the North Pole this year." This quote other than being absolutely ridiculous the wording makes no sense.

2008 June – National Geographic News – "We're actually projecting this year that the North Pole might be free of Ice for the first time in history". Thought it was ice free a million years ago!?? There goes National Geographic's credibility!

2008 June – Associated Press – Quoting NASA scientists – "In 5 to 10 years the Arctic will be free of ice in the summer". No scientists named, just NASA. Who are we supposed to trust? Seems that some of our "trusted" agencies are led by a bunch of fools!

2009 December – The non-scientist Al Gore – The Arctic Ocean may be nearly ice free in summer as early as 2014".

2012 – The Australian – "Enjoy snow now, by 2020 it will be gone". HUH!

2013 July – The Guardian – "Ice free Arctic in two years heralds' methane catastrophe". Say What?

2017 - The Sunday Morning Herald – "Snowy retreat: Climate Change puts Australia's Ski industry on a downhill slope".

2018 – A Harvard University Professor wrote – "The chance that there will be any permanent ice left in the Arctic after 2022 is essentially zero". Another Harvard educated idiot! There are currently 5.6 million square miles of ice in the Arctic.

2018 – Greta Thunberg – "Top climate scientists say: Humans will go extinct if we don't fix climate change by 2023". "A top scientist warning that climate change will wipe out all of humanity unless we stop using fossil fuels over the next five years". The post was deleted, after the five years came and went. Once again, a "Top Scientist" with no name.

2020 January - Glacier National Park – "They had to remove their signs that predicted its glaciers would be gone by 2020". Must have been embarrassing! Wonder where they stored the signs?

2021 December – LA Times – "A no snow California could come sooner than you think". This after several years of record snowfall in the Sierras.

2022 – Bloomberg – "The end of snow threatens to upend 76 million American lives". A few months later the Sierra Nevada mountains had the 2nd largest snowfall on record.

Further predictions we are now hearing is that the Atlantic Ocean will rise high enough by the year 2050 to reach the left arm elbow of Lady Liberty in the New York Harbor. Liberty Island is 15 feet above sea level (at high tide); the Statue of Liberty sits on a pedestal 151 feet high, the statue itself is 150 feet tall, the left elbow of the statue is about in the middle of the statue, so about 75 feet above the top of the pedestal. That said, the Atlantic Ocean would have to rise 241 feet by the year 2050 to reach the left elbow of the statue. That would equal a little more than 9.6 feet per year of sea level rise for the next 25 years. Does anyone in their right mind think that the Atlantic Ocean is going to rise by 241 feet in the next 25 years??? If that happens the city of New York will be no more. The states of New Jersey, Delaware, Massachusetts, Connecticut and any other states along the eastern seaboard will be no more. Further, while our oceans are independently named, they are all connected. One cannot rise by 241 feet without the others doing the same.

A new prediction is that the City of Boston is sinking because of warmer temperatures. Pray tell, how does a city sink from warmer temperatures? Not to worry it will be gone by 2050 due to the 241 feet of rising Atlantic waters. The new eastern seaboard will be somewhere around central Ohio. Buy your beach front Appalachian property now before the rush. Nothing of any consequence happened from the 1950's to beyond 2020 so it became necessary to move the goal post to 2050.

After all of the above "PREDICTIONS" and those listed are but a sampling of brilliant observations over the years. With the exception of the Atlantic Ocean rising 241 feet (that we have to wait for) have failed to come to pass which proves that this "Climate Change" movement is purely political. Our astute "Leadership" in the Biden white house wanted to eliminate fossil fuels and plunge our economy into something worse than the great depression and freeze or starve our citizens to death for absolutely bogus reasons. As outlined in chapter 3 the CO2 in the earth's atmosphere has not changed in centuries. They have no clue as to the National Security risk they are placing on the American people and that of our allies with their attacks on the oil and gas industry.

There have been so many ridiculous predictions from so-called "scientists", over- zealous no nothing politicians looking to pad their bank accounts, moronic celebrities and just plain idiots, (I think someone left the gate open at the Scientific Insane Asylum) NONE of those predictions have been realized. When will the American people wake up to the fact that these "scientists", et all, are nothing but scare mongers looking for notoriety having no factual basis for their claims. In doing so they are dead set on destroying America while filling their bank accounts and appeasing our enemies.

To quote another lead author of the above noted Declaration, "Professor Richard Lindzen, has called the current climate narrative "ABSURD", but acknowledged that trillions of dollars and the relentless propaganda from GRANT -- DEPENDENT academics and agenda driven journalists currently say it is not absurd".

The article goes on to say the cause in changing temperature is due to "Solar Cycles". "Solar cycles are the most devastating argument against man-made global warming".

An article "The Cold Truth Initiative" by a Climate Scientist Tom Luongo formerly of the University of Florida outlines the findings of "John Casey a former White House adviser, NASA consultant and a space shuttle engineer. He is now one of America's most successful climate change and climate prediction experts". The article entitled "Scientist Confesses: Global Warming is a $22 billion (per Year) Scam" . That equates to almost $42,000 per minute.

John Casey released this information in a document 164 pages long after reviewing the U.S. government studies where he found evidence that proves there is no human caused global warming. Documents that have been suppressed and held back from the general public in an effort to promote an agenda designed to control the American people.

Whatever global warming is occurring is natural due to Solar Cycles, sunspots, solar flares, polar shift, the earth's wobble all of which are natural occurrences. Mankind cannot prevent them any more than mankind can prevent the earth's rotation. Mother nature cannot be tampered with.

In Tom Luongo's article – "The first great lie: The World is getting Hotter…Fast" "The media jumps on "global warming" every time there is a heat wave or a hurricane hits the East Coast" "according to NASA's own data the world has warmed by .36 degrees Fahrenheit over the last

35 years. Measurement began in 1979". "The bulk of the warming occurred between 1979 and 1998. Temperatures have actually dropped since then. We haven't seen any measurable global warming in the last 21 years".

"Lie No. 2: The oceans are getting warmer" – "In 2007, as previously mentioned, Al Gore while accepting his Nobel Prize for his "global warming" Initiative (and quietly pocketing millions of dollars) Gore made a striking prediction" "The North Polar ice cap is falling off a cliff. It could be completely gone in summer in as little as seven years. Seven years from now (2007)". The Nobel Prize organization has absolutely no credibility! This is not their first atrocity; they gave Obama the Nobel Peace Prize 12 days after he took office. That one is still a head scratcher! Also, they gave the Peace Prize to Henry Kissinger for selling the South Vietnamese people into communism.

"The fact is that the North Pole Ice Cap is increasing in size". Reviewing NASA images of the North Pole between 2012 and 2013 you can see the ice cap has increased in size.

An article by Warner Todd Huston refers to A recent study in Norway concludes that "there is NO Man-Made Global Warming". Find it at "issuesandinsights.com the latest on global warming is there is no global warming". The study states that "Using theoretical arguments and statistical tests we find' the researchers say, "that the effect of man-made CO2 emissions does not appear to be strong enough to cause systematic changes in the temperature fluctuations during the last 200 years".

"Even if recent recorded temperature variations should turn out to deviate from previous variation patterns in a systematic way it is still a difficult challenge to establish how much of this change is due to increasing man-made emissions of CO2 and other greenhouse gases".

As issues and Insight added, "some high temp readings can be due to urban heat islands (UHI) where there is a lot of concrete and emissions all in one place. But it doesn't have that much effect on the whole world". No mention of reflective glass siding in the UHI scenario. Very much a contributing factor.

Warner states that "of course the zealots won't like the study. Well, they also won't like another new paper, this one from the University of Alabama in Huntsville climate scientists Roy Spencer and John Christy, who have submitted Urban Island Effects in Summer Surface Temperature Data, 1880-2015 to a science journal".

"They believe they have demonstrated that not only do the homogenized (adjusted) datasets not correct for the existence of the urban heat island (UHI) on temperature trends, the adjusted data appear to have even stronger UHI signature than the raw unadjusted data".

According to Spencer, "the bottom line is that an established 22% of the U.S. warming trend 1895 to 2023, is due to localized UHI effects, and that the effect is much larger in urban locations".

"The truth is there is no substantive proof for man-made climate change". Of course, climate changes. We know that for a fact. But to date it is all due to the planet's own process and man has contributed little to it". Warner Todd Huston on Facebook at: facebook.com/Warner. Todd. Huston.

In a recent video posted by Wide Awake Media, Geologist professor Ian Plimer:

"As soon as someone tells you it's warming, the reply you give is: since when"?

"We have been cooling down for the last 4000 years. It's all about when you start the measurements".

"If you take measurements from the Medieval Warming, we've cooled about 5 degrees since then. If you take measurements from the Roman Warming, we've cooled about 5 degrees since then".

I read an article the other day about an ice core that was taken back in the early 1990's from Greenland. It has been stored since then, I assume, in a freezer. Scientists have just gotten around to studying the core this past year. The article says that the scientists found a variety of plant specimens including a poppy seed in the ice core. Which means that at one point Greenland was not covered in ice and in fact had plant life at least in some areas. The scientists determined that sometime in the last 1 million years Greenland was ice free. This is not news, in past studies others have arrived at the same conclusion. This time, this group of scientists thought it was around 400,000 years ago. Immediately after making the statement that Greenland was ice free sometime within the last million years, one of them comes out and says "and if we don't start eliminating fossil fuels, we will see Greenland ice free again in the future". Okay, pray tell, what caused Greenland to be ice free 400 thousand years ago? Was it my SUV? These are the brilliant minds that the environmentalists are using to form their opinions. I wonder if any of these so-called scientists ever give any thought to what brings them to the sites to gather information for their "Scientific" studies. I would venture to guess their mode of transportation was not a horse and buggy!

A recent article at "worldorderreport.com" headline:" Scientists find no change in' Global Warming' rate since 1970." **TheDaily Sceptic** reports: A sensational science paper has blown holes in alarmist claims that global temperatures are surging. Just published results in *Nature* show "**limited evidence**" for a warming surge. "In most surface temperature time series, no change in the warming rate beyond the 1970s is detected despite the breaking record temperatures observed in 2023," the paper says. Written by an international group of mathematicians and scientists, it is unlikely to be acknowledged in the mainstream media where general hysteria reigns over the anomalous 2023 experience. As we have seen, constant misinformation is published to scare the general public and this is exemplified by climate comedy-turn Jim "jail the deniers" Dale forecasting almost daily Armageddon and exhorting people to "join up the dots".

If you need additional doomsday predictions visit the Farm Journal website AGWEB.com/doomsday. There is a wealth of info on failed predictions which include people starving to death in mass numbers.

It seems the more they try to prove their point that we are warming due to human activity the more they disprove it. The earth has been warming and cooling all by itself for millions of years. There is nothing mankind can do about it, no more than you can stop the sun from rising in the east.

Below is a late January 2025 article posted on Energy In Depth, one can only hope the madness is subsiding, perhaps the general population will wake up to all the skewed climate model reports and the bold face lies being told by some politicians and environmental zealots:

Why climate-change ideology is dying. Wall Street Journal. Momentous social movements begin to die the moment adherents figure out their leaders don't believe what they say. Liberal Protestantism's long decline started in the 1950s, when congregants began to wonder if their ministers still believed the old creeds (they didn't). Communism dies wherever it's tried because sooner or later the proletariat realize their self-appointed champions aren't particularly interested in equality. Many sects and cults dwindle the moment their supposedly ascetic leaders are revealed to be libertines. Something similar is happening to climate ideology.

Chapter 5
Wind Turbines / Solar Panels

An Opinion Article from the Wall Street Journal published in the "Energy In Depth" newsletter (IPAA)

Green electricity costs a bundle. WSJ. Opinion. "As nations use more and more supposedly cheap solar and wind power, a strange thing happens: Our power bills get more expensive. This exposes the environmentalist lie that renewables have already outmatched fossil fuels and that the 'green transition' is irreversible even under a second Trump administration."

Wind turbines require a very large volume of copper wiring to produce the electricity. Once again, the copper is mined from large strip mines around the world. One of the huge offshore windmills requires 4.7 TONS (about 9,500 lbs.) of copper to make the various components for ONE offshore wind mill. In addition to the environmental impact of the mining, crushing the ore, smelting (high heat) to extract the copper from the ore and transporting the copper to a facility that can make the wire required to construct the generator. The wire must then be transported to a wind turbine manufacturing facility to produce the necessary cables and windings for the generator. All of those processes are accomplished with the use of hydrocarbon sourced energy, primarily diesel fuel. Once constructed the wind turbine must be transported to the coast where it will be installed on the seafloor. That transportation will also include hydrocarbon fuels. The same process will be necessary for the construction and distribution of the rest of the wind turbine structure. The construction and installation of the turbine will include sea going vessels powered by diesel fuel. Most turbine blades are made of fiberglass-reinforced polyester or epoxy. All of these products are petroleum based. Many other parts of the turbine are made of steel. More manganese, iron, aluminum and other minerals. All the windmill components will have to be coated with a corrosion preventive chemical, likely derived from petroleum.

The installations of these wind turbines along the coast of the U.S. are killing whales and other sea life and having a negative impact on the commercial fishery. To add insult to injury, the contracts to build and install the wind turbines are held by foreign countries. British Petroleum (BP), Norwegian Equinor and the Dutch Marine contractor Heerema to name a few contractors slated to build offshore wind farms in the U.S. How does this benefit the U.S economy, construction monies going to foreign entities.

In addition to all of the environmental issues associated with offshore wind turbines there is a requirement for onshore substations to collect and distribute the produce power. These substations have to be located close to the shoreline and require undersea cable systems which also have a negative impact on sea life and require large amounts of copper.

What of Atlantic hurricanes? Will these turbines sustain hurricane force winds? How are they attached to the sea floor? What is the environmental impact of the installation? If drag force steel plates anchor the floating windmill, how much of the seafloor is damaged by this process? How much sea life is disrupted or killed?

Land based wind farms have a negative impact on the environment as well. Aside from the obvious visual detriment to the landscape. They are known to kill birds and they are prone to lightning strikes.

All of the above Windmills, land and sea require hydrocarbons to produce, transport and install! Tell me again how we can eliminate the use of fossil fuel!"

According to The U.S. Energy Information Administration (EIA), currently more than 60% of all U.S electricity is produced from fossil fuels. Natural gas being the largest producer at about 40 percent and coal the second largest at about 20 percent. Currently a little over 10% of the total U.S. power is produced from wind energy, approximately 3.4 percent is produced from solar panels. How many wind turbines and or solar panels will be required to offset the fossil fuel produced power? What will the landscape look like? How much copper will need to be mined to build wind turbines? What is the steel requirement for each turbine? How will copper and steel be produced without petroleum? How much polyester and/or epoxy will be needed for the blade construction? Wind turbine capacity is rated by what amount of kWh it can produce at maximum performance. Typically, the turbine operates at maximum capacity only about 10% of the time with varying levels of performance depending on wind conditions. Both wind turbines and solar panels are inefficient in that they are dependent on weather conditions. Solar panels are completely inactive at night and operate at about 50% capacity on cloudy days, wind turbines of course must have a windy day to operate efficiently. It is said that it takes 3 times the kWh capacity of wind turbines and solar panels to make up for kWh produced by coal or natural gas fired power plants. Check and see how much of your state's power generation capacity is hydrocarbon based; how many kWh do your source plants produce. Then multiply that by three to arrive at the necessary wind turbine and/or solar generation you will need to replace your power plants.

Solar panels are produced from a number of materials. Much like the EV battery the panels require more energy and fossil fuel burning equipment to mine, manufacture and transport than they save. The majority of large solar farms are located in the desert lands of the western U. S. The desert is a very windy and dusty place. Dust storms are frequent. To maintain efficiency the panels must be cleaned as dust accumulates on the surface of the panels decreasing efficiency. The cleaning requires a great amount of water, a commodity that is scarce in the desert. Further the panels have a life expectancy of 25 years and lose a percentage of their output annually. A 25 year old panel is likely not putting out much in the way of electricity. When looking at the size of most of the commercial solar farms in Arizona, Nevada and California the cost to replace depleted panels will be very high. Even at today's prices the cost for the panel replacement would be expensive. What will the cost be in 25 years? What will the resulting increase in price per kWh be? Where will you dispose of depleted panels?

Below is an excerpt from an article by: MICHAEL R. BLOOD Associated Press

January 30, 2025, 4:07 PM

Found at:

https://abcnews.go.com/US/wireStory/11-years-after-celebrated-opening-massive-solar-plant-118290949#:~:text=The%20Ivanpah%20solar%20power%20plant,compete%20with%20cheaper%20solar%20technologies.

"11 years after a celebrated opening, massive solar plant faces a bleak future in the Mojave Desert"

"What was once the largest solar power plant of its type in the world appears headed for closure just 11 years after opening, under pressure from cheaper green energy sources. Meanwhile, environmentalists continue to blame the Mojave Desert plant for killing thousands of birds and tortoises".

"The Ivanpah solar power plant formally opened in 2014 on roughly 5 square miles of federal land near the California-Nevada border. Though it was hailed at the time as a breakthrough moment for clean energy, its power has been struggling to compete with cheaper solar technologies".

Once dismantled, the older panels that are now depleted will have to go to landfills. Judging by the size of some of these solar farms (most are thousands of acres) the one mentioned above covers 5 square miles of dessert. There will have to be a sizable landfill to accommodate the waste. Just how environmentally friendly are solar panels?

A new headline at http://americanupdate.com:

"Biden-Subsidized Offshore Wind Developer Reports massive Losses in Latest Blow to Industry"

"Orsted, a Danish renewable energy group, contracted by the Biden Administration with subsides , estimated the value of its assets declined $472 million in the first half of 2024, largely due to delays at the onshore substation for its 704 megawatt (MW) "Revolution Wind" project off the coast of Rhode Island and Connecticut, the report showed the company also reported a variety of other problems, including losses related to its cancellation of its Ocean Wind projects in New Jersey and its abandoned effort to produce eco-friendly methanol in Sweden". "The Biden Administration has provided billions of dollars in subsidies to the offshore wind industry including Orsted". Yet they are still losing money! What is the true cost of these projects? Looks to me that Obama trained old Joe well!

The American Clean Power Association found there will only be 14 gigawatts of offshore wind production capacity deployed by 2030, less than half of Biden Admin. target. They didn't say at what cost to the taxpayer.

Another example of government reaction to a none starter. These projects are not thought through or evaluated, they just go with the

"Climate" therefore any expenditure, successful or otherwise is worth it. After all it's only your money!

An article from Warner Todd Huston in July 2022

(http://Ipatriot,comreport-wind-and-solar-has-added-nothing-to-the-power-grid.)

"Researcher David Stevenson recently wrote an essay detailing how little wind and solar power has been added to the power grid".

Stevenson of the Caesar Rodney institute Center of Energy & Environment, analyzed

(http://www.caesarrodney.org/pdfs/New_evidence_renewables_don't_reduce_ carbon_dioxide_emmissions2.pdf) "the effects of wind and solar and its effects on the nation's largest power grid between 2019 and 2020 and found that the addition of wind and solar was meaningless where it concerns mitigating the effects on global warming".

"The comparison of actual regional grid carbon dioxide (CO2) emissions between 2019 and 2021 shows increased use of wind and solar did not reduce emissions" Steenson wrote in his report.

"Wind and solar electric generation are actually poor technologies no one would use without permanent government mandates and subsidies and taxes that are adding $1 billion a year in power costs. They are also unreliable, non-recyclable, have negative environmental impacts, have shorter productive life spans than alternative power sources and take up a lot of ground. If it doesn't reduce carbon emissions, why are we using wind and solar"? He wrote.

"Wind and Solar are so unreliable, Stevenson noted, that other sources of power that the left claims are "dirty" must be cycled up and down over and over again to bridge the gap left when wind and solar fails".

"The lack of CO2 reduction by wind and solar comes at a high cost. Tax payers and electric customers provide expensive subsidies totaling almost $2 billion in 2020-21 period, or $1 billion a year".

"Billions of tax dollars are wasted every year on wind and solar, and no effect has been recorded to show it is worth what we've wasted on it".

There are all sorts of questions that arise relative to wind and solar. As mentioned earlier solar panels that are spread across the desert floor are susceptible to dust storms. A coating of dust on the panels can reduce efficiency as much as 50% depending on the thickness of the dust layer. The only way to remove the dust is to wash it off with water, something that is not readily available in the desert.

What of wind turbines offshore on the east coast, what happens in the winter when an ice storm hits? Do the blades get coated in ice? What damage does the added weight cause (ice is very heavy)? What of hurricane force winds, are these turbines capable of withstanding a force 5 hurricane or do we see the sea floor and beaches littered with wind turbine debris? I'm sure manufacturers and installation companies will swear up and down that the turbines can withstand anything that mother nature can throw at them, according to their computer models.

Computer models are unreliable when it comes to duplicating natural events! Why are wind turbines killing whales? It is said that the excessive vibrations from the turbine blades create high frequency sound waves that cause whales to be disoriented and disrupts their brain waves which causes abnormal behavior which in some cases causes death. There are so many unknowns related to this technology it seems once again that government entities are putting the cart before the horse and the cart is loaded with your tax dollars.

An article at https:// ipatriot.com/Study-bucks-lefty-claims-finds-offshore-wind-farms-are-bad-for-marine-life/

"Leftist eco warriors claim that offshore wind farms are green ideals, but a new study from Germany finds that these farms are bad for marine life despite what leftists claim".

Per Town Hall

(https://townhall.com/tipsheet/spencerbrown/2022/11/30/study-finds-green-offshore-wind-farms-have-substantial-impact-on-ecosystems-n2616532):

The researchers looked at one area and ecosystem in particular – North Sea – where the world's largest offshore wind farm (OWF) opened earlier in 2022. Relying on modeling and simulations in order to see what the "systematic, large scale, time integrated response of ocean to large OWF clusters would be, the study authors found evidence that offshore wind power generation, supposedly great for the planet, actually introduced factors that negatively impact sea life".

While it seems logical that wind farms will disrupt sea life. Once again, the "researchers'' are using computer models to generate results which leaves me skeptical regarding the outcome of the study.

An article in https://www.libertybugle.com/beware -100-green-energy-could-actually-destroy-the-planet/

"According to "Foreign Policy" analysis, moving to a "carbon free" energy future "requires massive amounts of energy, not to mention the extraction of minerals and metal at great environmental and social costs"

"Here are some numbers. Going all in on batteries, solar, and wind require":

"34 million metric tons of copper, 50 million tons of zinc, 40 million tons of lead, 162 million tons of aluminum and 4.8 BILLION tons of iron". All of which must be extracted, processed, and transported using diesel fuel. Just how environmentally friendly is that?? Eliminate petroleum and all of these projects are dead in the water, some quite literally.

Chapter 6

Sources of Electricity Generation in the U.S - the Power Grid

To point out a few states that are dependent on fossil fuels for power generation.The state of Michigan receives about 40% of its power from Coal powered generation facilities. Pennsylvania gets more than 52% of its power from natural gas. The state of Utah gets 53% of its power from coal fired facilities while an additional 29% comes from natural gas facilities. Almost 50% (47.5 %) of California's power comes from natural gas. New York is 47% dependent on natural gas for its electricity. 77% of the power for Hawaii comes from petroleum fired generating equipment. The majority of Colorado's power, 37% comes from coal fired generators with an additional 29% comes from natural gas facilities. Nebraska depends on coal for 49% of its power. The state of Mississippi is 76% dependent on natural gas for electric power generation.

Modern technology has provided power generation equipment that is highly efficient and environmentally friendly. Coal fired plants operate on what is known as "Clean Coal" this is not the coal technology that provided power to the Pittsburgh steel mills of the past. New environmental rules and regulations have forced innovation in technology to drastically reduce emissions from both coal and natural gas fired generation equipment. Hydrogen is probably the cleanest fuel on earth but natural gas is likely the second cleanest burning fuel on the planet. Methane (natural gas) is one carbon element from pure hydrogen. Natural gas generation plants are more environmentally friendly than either solar or wind power generation. The footprint for a natural gas generation facility is most certainly a great deal smaller than the Real Estate necessary for a windmill farm or an array of solar panels (never seen a natural gas fired power plant cover 5 square miles of desert land) and there are no disposal or recycling issues with a natural gas generator.

The point here is that many states depend on fossil fuel to keep the state operational from household utilities to complex commercial, industrial and manufacturing facilities as well as state and federal government facilities. This is not something you just turn off at the switch and say we are going to use other means to generate the nation's electrical power. Solar and wind power are not going to compete with fossil fuels. If there were enough real estate available to accommodate those facilities it would completely destroy the environment across the country. The fisheries offshore on both coasts will experience severe damage not to mention the scenery.

Below is a list of the sources of electrical power generation in the U.S. (as published by U.S Energy Information Administration, EIA) the highest percentage (60.4%) of power generation equipment uses fossil fuels in one form or another. There is no practical way to replace fossil fuel power generation with any other means of generation, save nuclear energy and we know the risk and opposition associated with that.

You can create an environmental catastrophe by placing solar panels and or wind turbines all across the deserts of Nevada, California, Arizona and New Mexico and still not be able to duplicate what is being produced from fossil fuels. If you were to attempt such an environmental nightmare you would kill countless species of wildlife from the desert tortoise, coyotes, roadrunners, jack rabbits, and many others. In addition, solar panels are produced from a variety of sources, petroleum being one ingredient.

An article in the publication "Power the Future" entitled "Where Green Meets Red". The article states that the Biden administration, push to electrify nearly every facet of everyday life in America"- power plants, cars, the heating and cooling systems in homes and buildings- is dangerously increasing our dependence on China and is jeopardizing our nation's security".

Once again using California as an example, in 2022 there were approximately 14.2 million vehicles registered in the state about 5,000 were electric vehicles. California has over the last several years encountered rolling brownouts and blackouts. The governor in September 2022 asked the owners of Electric vehicles not to charge them because it would over stress the grid and cause brownouts and blackouts in many areas of the state. What happens when you increase the number of EV's to 20,000? The U.S as a whole, registers about 283 million cars. Where is the power going to come from to charge that many vehicles? Answer: There won't be enough power or the need for it because you won't have a car. The cost of the car and then the cost to maintain and charge that vehicle will be well beyond affordable for the average person. The sad part is that the cost of power will go up exponentially due to electric vehicles and the cost of power for your home will go along with it. There will be a breaking point to where people will have to decide to turn off the lights or the home heating system.

Below is an excerpt from the publication "Energy in Depth"(IPAA):

"Whole 'Nother Rent': Soaring Electricity Prices In Kamala Harris' Home State Are Crushing Residents. Daily Caller. California's electricity prices are surging, burdening residents with bills that in some cases exceed the rent they pay for their homes or businesses. Prices have increased thanks in large part to the state's push to make infrastructure upgrades like burying power lines, as well as demand increases driven by EV adoption and the proliferation of power-hungry data centers needed to sustain the growing artificial intelligence (AI) sector, according to the WSJ. These costs, as well as the capital costs of new wind and solar farms that policymakers want to replace fossil fuel generation for a green grid, are being passed on to California's consumers, leaving ordinary people on the hook for huge electricity bills."

An article, published by Rightwing.Org, called the "Electric Car CRISIS" analyzes the problem of forcing EV's on the general public. The article poses the question, "are environmentalists and their friends in Congress moving too quickly to force America onto green energy"? "Across the country, energy grids are struggling to keep up with current demand". "In Texas alone, the state's energy distribution company

(ERCOT) is asking residents to be mindful of their energy usage to avoid rolling blackouts, which have become more common in recent years. The question is, can renewable energy replace fossil fuels"?

"Whether it can or not, auto manufacturers are moving quickly to transition cars off of gasoline and onto battery power. Still, the amount of electricity to charge millions of cars is phenomenal, and the growth in electric vehicle usage could present a unique challenge to each state's grid". Update: Some auto manufacturers are moving away from EV's due to huge losses per vehicle and because they can't sell them.

The necessary infrastructure to change to an all-electric fleet in the U.S is an impossibility. The primary reason is getting the necessary infrastructure to accommodate new power sources. High tension systems don't just appear overnight. There are years of development that have to come first. The Rights of Way, if not on federal land, will have to be acquired which assumes land owners will be willing to allow utilities to use their land for their needs. If land owners are not willing to allow the utilities access to the land, then they will have to try and find an alternative route. Rights of Way are not a simple part of new construction; it could take years.

In addition, there will have to be environmental impact studies done along the proposed Rights of Way. Those studies can also take an extended amount of time (beyond reasonable) especially if the Right of Way interferes with "Habitat".

While all of that is going on there will have to be engineering studies to determine the size of the project, demand forecast relative to total cost. What will be the cost per kWh when the project is complete? While the utility company will get tax benefits from the government, they will also have to establish a rate base for the sale of the power. The price per kWh you are accustomed to will no longer apply. The billions of dollars necessary to build the infrastructure will have to be recouped in a reasonable amount of time. The power you receive now is delivered through decades old equipment which was built for far less money than what will be required today.

The below article from the IPAA publication “Energy In Depth” shows the result of actions by uninformed agenda driven state legislators and overzealous environmentalists:

California's 'ambitious' policies to reduce greenhouse gases drive a surge in electricity costs. Fox Business. Electricity prices in California have surged significantly over the past several years, which officials say is partly due to green energy policies being pushed throughout the state. The Legislative Analyst Office (LAO) released a new report revealing that monthly electricity rates for residential customers in California have increased by an average of nearly 50% since 2019. "In general, average residential electricity rates in California have grown faster than inflation in recent years, rising by about 47% over the four-year period from 2019 through 2023, compared to overall growth in prices of about 18%," the report reads.

It seems that there is no thought that goes to the impact of these decisions to alter a system that works just fine as it is.

What is U.S. electricity generation by energy source?

In 2023, about 4,178 billion kilowatt hours (kWh) (or about 4.18 trillion kWh) of electricity were generated at utility-scale electricity generation facilities in the United States.[1] About 60% of this electricity generation was from fossil fuels—coal, natural gas, petroleum, and other gases. About 19% was from nuclear energy, and about 21% was from renewable energy sources.

The U.S. Energy Information Administration estimates that an additional 73.62 billion kWh of electricity generation was from small-scale solar photovoltaic systems in 2023.[2]

U.S. utility-scale electricity generation by source, amount, and share of total in 2023

Energy source	Billion kWh	Share of total
Total - all sources	4,178	
Fossil fuels (total)	2,505	60.0%
Natural gas	1,802	43.1%

Energy source	Billion kWh	Share of total
Coal	675	16.2%
Petroleum (total)	16	0.4%
Petroleum liquids	12	0.3%
Petroleum coke	5	0.1%
Other gases[3]	11	0.3%
Nuclear	775	18.6%
Renewables (total)	894	21.4%
Wind	425	10.2%
Hydropower	240	5.7%
Solar (total)	165	3.9%
Photovoltaic	162	3.9%
Solar thermal	3	0.1%

Energy source	Billion kWh	Share of total
Biomass (total)	47	1.1%
Wood	31	0.8%
Landfill gas	8	0.2%
Municipal solid waste (biogenic)	6	0.1%
Other biomass waste	2	0.1%
Geothermal	16	0.4%
Pumped storage hydropower[4]	-6	-0.1%
Other sources[5]	10	0.2%

Data source: U.S. Energy Information Administration, *Electric Power Monthly*, February 2024; preliminary data

Chapter 7
Government Actions and Proposals and Their Impact

The "Green New Deal" (somewhere in the title the word SHAFT should appear) an idea brought forth by a New York City Marxist bartender turned politician. It is blatantly obvious she has no clue of what the proposal does to the people and the economy of the country. Nor does Biden who jumped right on board and issued edicts to any government agency involved in energy. This is not about the climate, it is about CONTROL.

In line with the above proposed "deal" the Biden administration, in 2021, provided, $7.5 "BILLION" for charging stations to be built across the county. They are expecting 500,000 stations to be built and operational by 2030. Three years into the project, to date, they have completed 8 charging stations; it is unknown how many are operational. It took 3 years to get 8 facilities built. Is there anyone on the planet that believes they will construct another 499,992 in the next 6 years??? This waste of taxpayer funds has the same odor as Obama's Solondra deal, only on a much larger scale. The Solandra deal is where a group of Obama supporters were given $535 million to make solar panels. The deal resulted in Solondra losing $528 million and then declaring bankruptcy while the principals' received bonuses and Obama received large campaign donations. It is estimated that each of Biden's charging stations will cost about $15,000 which says that about $120,000 has been spent. Where is the other $7,499,880,000???

In May of 2023 Senator John Kennedy of Louisianna, at a Senate hearing questioned David Turk, the Deputy Secretary of the Department of Energy, under Biden. His questions included "how much will it cost the U.S taxpayer for the country to become carbon neutral"? Mr. Turk could not provide a definitive answer. Sen. Kennedy estimated it would cost $50 trillion to be carbon neutral. With that number on the table, Sen. Kennedy then asked "if the American tax payer spent $50 trillion for carbon neutrality by the year 2050, how much would that reduce world temperatures"? The question was asked several times and the Deputy Secretary could not answer. During his testimony Mr. Turk answered the question regarding the timing for the U.S. to be carbon neutral he referred to his sources of climate information with the below quote:

"So, I think, according to the climate scientists around the world, and certainly the "cutting-edge" scientists that we need to rely on here in the U.S., we've got to get carbon neutral by 2050, and I'm very comfortable with that target, and I think that's the appropriate target—" Note no "Scientists" named.

Yet he could not identify the worldwide benefit of the U.S becoming carbon neutral. According to the Deputy Secretary of the Department of Energy the U.S is responsible for 13% of worldwide CO2 emissions. It is unclear if that percentage includes natural events or if it is attributed solely to human activity. Once again this assumes that human activity is not natural for life on earth. Let's assume it is all inclusive. If so, 96% of the 13% occurs naturally from active volcanoes, seafloor expansion, wetlands, etc. things you will not change. 96% of 13% comes to 12.48% of emissions in the U.S. that come from natural causes leaving 0.52% due to human activity. The atmosphere contains >.40% CO2. The .52% of CO2 attributed to U.S human activity would be .52% of .40% or 0.002 % of the total atmospheric CO2 content. That would be a U.S. nationwide number and assumes that all human released CO2 rises to the Atmosphere, IT DOES NOT! For the most part it is plant food!

Now let's take a look at California. California is about 10% of the nation's population so it stands to reason that their emissions would be about 10% of the national emissions of 0.002% for all state activity which comes to 0.000002% of the U.S. contribution to the global atmospheric CO2. Eliminating the California and/or the national number of CO2 is not going to reduce global CO2 atmospheric content. California nor U.S. carbon neutrality will reduce Global temperatures one iota.

Let's assume now that the deputy Secretary meant that the 13% was all human activity, nothing natural. The result would be about .052% of the .40% total atmospheric CO2 content if in fact all of the 13% went directly to the atmosphere which we know it does not. Most of it will be absorbed by plant life. What minute amount that may rise to the atmosphere is not going to create any global catastrophes nor will eliminating it prevent natural emergencies. Carbon neutrality is an ABSOLUTE IMPOSSIBILITY unless the population stops breathing. That would include the Deputy Secretary of the Department of Energy.

I would maintain that it is impossible for any country to become carbon neutral since the majority of carbon emissions worldwide comes from natural sources. As mentioned earlier 96% of the carbon in our atmosphere comes from natural events such as volcanoes, seafloor expansion, wetlands, animal life, etc. not only is it impossible to accomplish carbon neutrality but it would be devastating to plant life. All plant life depends on CO2 to exist. In turn the plants produce oxygen for us to breathe.

Once again, we hear "Scientist Say" and "They Say" without any specific mention of who "they" are, their credentials, or their research findings by specific cases. Those specific research findings would have to be accomplished without bias computer modeling. Chapter 4 outlines what true climate "scientists say". The Deputy Secretary speaks of "cutting edge scientists" he depends on here in the U.S. yet there is no mention of any scientists by name, company, department or organization. Once again if they rely upon government funds for their livelihood, I would disregard most if not all of their statements.

In addition, there are now ridiculous proposals for regulations to outlaw natural gas stoves, water heaters and other appliances that require natural gas. Natural gas is the cleanest burning fossil fuel on earth with the exception of Hydrogen. Again, the symbol for natural gas (methane) is CH4 which translates to one carbon atom to four hydrogen atoms. This means that methane is one carbon atom from being pure hydrogen, hydrogen being the cleanest fuel in the world. The exhaust from a hydrogen powered engine (a fuel cell) is water vapor. Environmental activists have been pushing for Hydrogen fueled vehicles for years without realizing or caring about the consequences of the conversion. What they don't seem to consider is that of the 2% of the atmosphere being the global blanket of "greenhouse gases", 1.95% is water vapor. Therefore, they want to eliminate CO2 (which cannot be done) and replace it with water vapor. What would the greenhouse gas blanket look like if the world fleet were to be all Hydrogen powered vehicles all of which would emit water vapor? How large above the current 2% of the atmosphere would that blanket grow. You can bet there would be a lot more rain! Further, coming back to earth, what would the roadways in Minnesota look like in January with water vapor emissions from all vehicles. The vehicle emission would turn to ice crystals immediately creating extremely hazardous road conditions. How about the humidity in the southern states during the summer months, would you still be able to breathe? All the people killed by this conversion would just be considered collateral damage by the environmental zealots.

According to an article in Energy in Depth (IPAA) Now the state of California is proposing that all gas stoves come with a tobacco style health warning label. The state is claiming that the "blue flame releases air pollution in your kitchen". Talk about dealing in minutia. "California legislators have wanted to pass a bill that would require such warning labels on gas stoves for sale in stores and online". Don't know why they would think anyone would read the label and of course there is no consideration for the added cost to the industry or the consumer. Kind of like the label on your mattress.

Another brilliant proposal comes from California. No more new internal combustion engine vehicles sold from 2035 forward. This mandate, aside from being unconstitutional, forces the sale/purchase of electric vehicles, likely to unwilling buyers. No consideration of the fact that California's power grid cannot handle an influx of that many EV's. My guess is the used Internal combustion engine car market will be very strong in 2035. This of course assumes you can find a gas station. Absolutely no consideration for the U.S. white and blue-collar jobs that will be eliminated due to this moronic edict. The majority of batteries and EV's will be imported from foreign countries.

To assist in attaining the EV 2035 target date, the California Energy Commission has proposed the takeover of all California oil refineries. They claim the intent of the proposal is to reduce the chance of price gouging at the pump. Have you ever known a government entity to operate efficiently enough to reduce or control the price of anything? Refiners make between $.16 and $.25 per gallon of gasoline, the state of California makes about $1.40 just who is it that's gouging? The real intent of this proposal is to phase out oil refining to accommodate the 2035 EV mandate. They would slowly shut down refineries to short the market of fuel, closing gas stations statewide. If you can't buy gas you have to go to electric vehicles. My guess is that most people would leave the state. One motivator would be the lack of food in the grocery stores as farmers and truckers would not be able to buy diesel in the state. They won't be able to tend to their farms or move their produce to market. Total economic collapse! Also, not a consideration is that the state of Nevada is 90% dependent on California for its motor fuels and Arizona is about 60% dependent on California refineries for its motor fuels. Shut down California refineries and you shut down the California economy and that of two additional states.

According to an article in the Energy in Depth (IPAA) newsletter "California Republican Assembly Member Joe Patterson introduced a bill to delay an update to California's low carbon fuel standard that the state has estimated could add 65 to 85 cents per gallon to the price of gasoline as soon as January (2025). The proposal would bar the California Air Resource Board from updating the program, which is a cornerstone of state efforts to reduce greenhouse gas emissions until January 2026". This bill is introduced by a republican and does not stand a chance in the California super majority democrat assembly and senate. To them it is not about the well-being of the people it's about power and control. Update: the bill proposing the Low Carbon Fuel mandate just passed both assembly and senate and has been signed by governor gruesome. Assembly member Joe Patterson's bill was completely ignored. Therefore, the price of gasoline in California will increase by about $.60 per gallon as soon as January 2025.

California over control proposals placed on the Oil refiners in the state has now caused one major refiner to make the decision to shutter their LA plant which reduces California's motor fuel supply by about 8% The bureaucrat's constant rulemaking has put other refiners in a position of following the lead of the first refiner and may shut down their plants as well. THEN WHAT? News flash: Chevron has introduced the possibility of exiting the California market. It is unclear if they are talking about shutting down their two refineries in California or selling them. Shutting down those two refineries would reduce the California refining capacity by about 32%. Perhaps this is an opportunity for the California Energy Commission to step in to the refining business. God help us!

California 2035 mandates will require massive changes to the existing power grid. In order to expand the power grid, to accommodate EV's, each expansion and/or new installation project will require an Environmental Impact Study, God forbid they find a kangaroo rat habitat or some endangered weed along a proposed right of way of any given project. Given the bureaucratic inefficiencies that will add about 2 to 3 years to the project. When these decisions are made, politicians just shoot from the hip. They think they are saving the planet which is an absurd assumption and will likely have the opposite effect. Mankind cannot and will not change the earth's climate regardless of what politicians think they have the power to do. Their motivation is pure greed for power over the people and self- enrichment as well as a complete waste of taxpayer dollars.

The Biden administration gave directives to the EPA to force heavy truck conversion to electric vehicles. The EPA which consists of appointed bureaucrats not elected officials. The EPA is probably the most dangerous agency in the federal government. These people for the most part have never held a private sector job. They sit around a table and discuss what they perceive to be necessary to protect the planet. They appear to give little concern relative to what their actions do to the country overall with regard to the economy and the everyday lives of the people. If the EPA gets what they want, millions of Americans will lose their jobs while experiencing an extreme increase in the cost of their food, utilities, transportation, clothing, and their entire livelihood. As explained in Chapter 1 "Nothing Moves Without a Truck". The entire supply chain is at risk. In addition to heavy duty trucks, they are also going after farm equipment and locomotive engines.

Typically, a farm tractor will go into the field with a tank of diesel fuel that will last the whole day. How long will an electric tractor last in the field??? Will it require charging after a couple hours? If so, how long will it take the tractor to get back in the field? Will it take 3 or 4 hours to charge? The task for the day will determine how long the tractor will be able to stay in the field? Some farm activities require more power / resources than others. Deep rip plowing requires more power than seeding. What happens if the electric tractor runs out of power in the field? Do you dispatch a mobile power generator to the field to spend several hours charging the tractor? Would the generator be diesel powered or a series of batteries? If diesel powered, where do you get the fuel to run that equipment? Diesel tractors are multi-task vehicles and the most efficient. They are used to deep plow fields, prep the field for seed, discing, cultivating, spraying, fertilizing, harvesting, many farm activities. In addition to tractors, they want to convert all diesel-powered farm equipment to electric power. This would include harvesters for most produce, potatoes, tomatoes, wheat, cotton, carrots, corn, grapes, and many others. You can include nut tree shakers which are used to harvest nuts and some fruit.

Limiting the time a tractor can stay in the field limits the farm's production. Converting to electric vehicles would require the farmer to add more very expensive equipment and personnel. Where one tractor is required now it would likely have to be replaced with three at a cost far greater than the diesel equipped alternative. Just the inconvenience of EV's adds to the cost of the operation and the cost of the product. Add the inefficiency of operating EV produce haulers and you can count on food prices tripling at a minimum and count on supply chain disruptions. Your grocery store shelves will be empty!

To demonstrate how dangerous the EPA is, a recent Energy In Depth Article addresses EPA methane rule changes:

EPA faces methane rule challenges from states, industry groups. Bloomberg Law. Dozens of states and trade associations in separate suits are seeking review from the D.C. Circuit of the Environmental Protection Agency's latest methane rule. The rule "exceeds the agency's statutory authority and otherwise is arbitrary, capricious, an abuse of discretion, and not in accordance with law," the **Independent Petroleum Association of America**, Texas, and others said in their suits filed with the US Court of Appeals for the D.C. Circuit on Wednesday and Thursday. Both suits are asking the court to declare the rule unlawful and vacate it.

It appears they just make up RULES and float them out as a mandate to see what might happen. Lawsuit after Lawsuit result in "Rules" being made beyond the authority level of the agency. The lawsuits take a great deal of tax payer money away from issues that are of far greater importance than the ego of an EPA operative.

Once again, the EPA is a very dangerous government agency. As mentioned, it is run by bureaucrats with little oversight. They receive an idea from a political entity and then act on it. The EPA has launched an all-out war on the energy industry and the large portion of the population that rely on natural gas appliances. They are accomplishing this in the form of methane regulations. As stated in Chapter 3, the methane content in the atmosphere is .12% and has remained that way for well over a century. Methane is not causing a rise in global temperatures, if so, let's drill the seeps (mentioned in Chapter 3). Once again this is about control of the people and industry not global warming/climate change or a climate emergency, whichever label you want to put on it. It is about a political movement led by Marxists, ignorant or stupid radicals and complete idiots who have no idea of how much they depend on oil production in their everyday existence. Their goal is simply to destroy the United States. The movement is exacerbated by politicians who see opportunity for power and control in taking advantage of popular "opinion" regardless of fact.

None of these nonsensical regulations will change the content or concentration of any of the existing gases in the atmosphere. The regulations do far more harm than good. Mother Earth is going to do whatever she wants regardless of government rules and regulations.

As an example of the hypocrisy associated with these rules and regulations, the Obama administration committed the U.S. to the "Paris Climate Agreement." This so-called agreement is financed by the member nations (mostly financed by the U.S.) who set their own emission targets and must report their success or failure to attain those emission levels to those who administer the rules. China stepped up to sign on to the agreement as well. In the negotiations for setting the standards China was given a pass on following the proposed emission standards until 2030. While the U.S. has set unrealistic goals to reduce emissions now while China is building coal fired power plants (over 300 coal plants) since the signing of the agreement. Understand one thing, when the year 2030 rolls around China will renege on their promise to reduce emissions, while the U.S political machine continues to destroy our economy.

As an example In an article published by

http://americanactionnews.com/featured 2024/10/10 "The Biden - Harris administration has doled out billions of dollars to a left-wing nonprofits who's CEO called for the end of carbon emitting companies and for the U.S. to engage in a "world leading mission" with the Chinese Communist Party (CCP) to phase out fossil fuels." That would be the United States fossil fuels not China. Another example of the total ignorance of politicians and the absolute corruption of those they are associated with.

The article continues with: "We're engaged in the world leading mission, or we hope, with China, of changing the way that everything is organized in society, the way that heating and lighting and air conditioning and transportation take place," CEO of the Coalition for Green Capital(CGC), Reed Hundt, said during a CGC webinar held on Aug. 23, 2024 that outlined how the nonprofit plans to spend the massive grant award. A week prior to the webinar, the EPA awarded the

CGC $5.1 billion under the Inflation Reduction Act's (IRA) Greenhouse Gas Reduction Fund, a program that Republican lawmakers have warned could potentially fund entities with ties to the CCP." Guaranteed the money doled out to this organization will end up in the pockets of the officers of the company. Did the EPA ask for any information regarding a budget or accounting for expenditures? Doubtful!

Biden also proposed that he give third world nations "reparations' using U.S taxpayer dollars due to the damage caused by "Climate Change." One, how is that our responsibility; Two, what damage are we talking about; Three, how can anyone tie any "Damage" to "Climate Change" when the 35 year temperature change is a whopping .36 degrees Fahrenheit. The whole premise of this "Theory" is untenable and has the unmistakable odor of government corruption.

The question has been posed, on several occasions, from several sources "Can the Country recover from the Biden-Harris administration?" One can only hope!

The administration has no money for U.S hurricane victims but they have plenty of money to give to organizations like CGC that are hell-bent on destroying our nation. As noted above, China has no intention of living up to their "Paris Climate Agreement" promise of reducing carbon emissions after 2030. They will continue to operate their existing coal fired power plants and likely build more. They are currently moving forward to position themselves as the world leader in manufacturing and operating under the guise of future cooperation. It's

"TRUST ME "we will conform when the time comes, by that time they may well be in full control.

The Biden administration ban on drilling in Arctic National Wildlife Refuge (ANWR). The proposed drill site on the tundra is about 2,000 acres. ANWR consists of 19.6 million Acres, the drill site equals 0.0001% of the refuge. If you compare the 2,000 acre drill site to the size of the tundra it would be about the size of a postage stamp on a large shopping mall parking lot. The drill site has an estimated 10 billion barrels of recoverable oil reserves as well as large volumes of natural gas. This would reduce the U.S. dependence on foreign oil especially California's dependence. While the nation as a whole can become energy independent with the proven reserves in much of the lower 48 states, California is still 75+% dependent on foreign oil, mostly due to state government intervention engaging in their war on the oil industry. To maintain the existing refiner production in California, refiners are forced to purchase foreign oil to replace the declining domestic production due to state regulations relative to producing oil. The dependence on foreign oil creates trade deficits and has a negative impact on the California employment rate as well as state taxes. Purchases of foreign oil sends money overseas, for the most part, to countries that don't like America. The use of domestic production provides jobs for California residents and state tax revenues increase from the taxes paid by the employed as well as the producing companies. The money stays in our country.

When the ANWR reserves were first discovered, congress was up in arms about drilling. Some zealots made presentations that were absolutely false about the environmental impact of producing oil at the site. They put up pictures of Denali National Park and a variety of wildlife pictures claiming the drilling would destroy the wildlife and their habitat. Pristine rivers would be polluted and several other doomsday predictions.

If you have ever been to the arctic circle, (I have) and viewed the landscape it is as flat as a fritter. The tundra extends for hundreds of miles. The primary wildlife other than sea birds is the caribou herd, possibly a scattering of polar bears and billions upon billions of mosquitos. When the Trans-Alaska pipeline was being built the so-called environmentalists protested with claims that the construction would destroy the caribou herd making them extinct. At the time the herd was about 6,000 animals. About 5 years after the pipeline was finished and operational the herd had grown to over 15,000 animals. Why??? The pipeline is for the most part above ground and is a heated system. Though it is insulated it still emits heat. The mother caribou would calf next to the pipeline and the heat would provide protection for the calf from the extreme winter temperatures, preventing the calf from freezing to death. In addition, the living quarters for the employees that worked on the north slope drill sites were raised about 6' above the ground. The caribou would and still do lay under those buildings for protection from the elements.

There is a local community of about 300 people near the proposed ANWR drill site who were looking forward to the facility becoming operational. The oil field would provide an economic boost to their community by providing jobs for the residents. Oil workers are paid very well especially in remote areas. The operation would also provide a near-by source of fuel. Each of the drill sites on the north slope have a small operational refinery to fuel the vehicles for onsite maintenance activities. Banning drilling of the ANWR site is detrimental first to the Alaskan economy as well as the U.S overall. Allowing ANWR production would reduce California's dependence on foreign oil (currently 75+%) by about 90% and substantially lower the U.S. trade deficit. Allowing additional California production would entirely eliminate the need to purchase foreign oil, save some highly specialized feedstocks for specialty products.

Incidentally you can forget about converting north slope vehicles to EV's. As it is extremely difficult to charge an EV in cold weather.

The latest Biden blunder is to outlaw drilling on both the Atlantic and Pacific coasts. Below are some articles found in the Energy In Depth (IPAA) publication:

Biden's climate narcissism could destroy America's energy advantage. The Hill. Opinion. President Biden, on his way out the door, has permanently put over 625 million acres of federal waters off-limits to oil and gas exploration. This is the move of a spiteful old man eager to undercut one of the highest priorities of his successor, President-elect Donald Trump: U.S. energy development. Reducing acreage available for oil and gas exploration is a profoundly ignorant and damaging measure.

Biden bans new oil and gas drilling along most U.S. Coasts. New York Times. President Biden announced on Monday what he called a permanent stop to new oil and gas drilling across more than 625 million acres of U.S. coastal waters, saying he was making the move because drilling posed unnecessary risks to the environment, public health and the coastal communities' economies. Ron Neal, the chairman of the offshore committee of the **Independent Petroleum Association of America**, an oil and gas trade group, called the ban "significant and catastrophic." CNN, E&E News, Offshore Magazine, and World Oil also report.

Biden Offshore Drilling Ban "Significant and Catastrophic." Independent Petroleum Association of America. **Independent Petroleum Association of America (IPAA)** Offshore Committee Chairman Ron Neal, the President of Houston Energy L.P. and CEO of HEQ Deepwater, issued the following statement in response to the announcement of the Biden Administration's plan to remove millions of acres of U.S. oceans from potential oil and gas drilling, barring the sale of new oil and gas leases along the Atlantic coast, the Pacific coast, the eastern Gulf of Mexico and portions of the northern Bering Sea: "President Biden's decision to ban new offshore oil and natural gas development across approximately 625 million acres of U.S. coastal and offshore waters is significant and catastrophic."

Biden administration creates final limits for oil drillers in Alaska. NYT. The Biden administration announced new protections for 1.3 million acres in the North Slope of Alaska, a final effort to shield it from oil companies eager to drill in the ecologically sensitive Arctic environment. President-elect Donald J. Trump returns to the White House in 2025, and he has pledged to grant fossil fuel companies broad access to American land and federal waters.

In addition to the ban on drilling in coastal regions he is banning drilling in Nevada's Ruby Mountains. The Biden administration has been the single most anti-oil administration in the history of the nation. They have demonstrated a total disregard for the people of our country as well as our "well-being". The entire United States economy revolves around hydrocarbon fuels, oil, natural gas and coal. If you eliminate those fuel sources you destroy the economy which is precisely what the Biden administration had intended to do since 2020. Without oil there is no supply chain, without natural gas and coal there is very limited electricity, save the area specific hydro power and the minutia provided by wind and solar which are also not possible without petroleum.

Another blunder of the Biden administration, on the way out the door, he has banned natural gas water heaters, an appliance that millions of Americans depend on for good hygiene. A prime example of government over-reach. He has done everything he possibly could to make the transition of power as painful as possible. This will give the Trump Administration a little heartburn but the people of the U.S are the ones that will suffer the most until these wrongs can be righted.

The Biden administration also wanted to outlaw "Fracking" (hydraulic fracturing), The mantra is that "Fracking is a health hazard and contaminates groundwater". No truth in any of the claims. Once again statements are made by politicians parroting anti-oil activists. Ground water does not get contaminated from fracking an oil or gas well. Google the anatomy of an oil or gas well and look at the construction diagrams (Illustrations). You will see that the well is drilled a few hundred feet below the ground water table and then cemented to prevent water from entering the well bore and oil / gas from entering the water. It is completely sealed. The well is then extended to the oil or gas reservoir. Fracking is a technique that has been used in the oil and gas industry for more than 75 years. Why is it all of a sudden, a danger to the population and the water supply? The difference in Fracking today versus decades ago is that the process is used to stimulate horizontal wells as opposed to vertical wells. This requires more equipment and the process is a bit more costly and time consuming. How hydraulic fracturing works is fluid made up of water, sand and some chemicals, (Fracking companies have proprietary recipes for their fracking fluids) Some use dish soap as a lubricant (chemicals). The fluid is injected into the well at very high pressure which fractures the oil or gas bearing rock. The sand in the fluid is forced into the fractures in the rock and holds them open to allow oil or gas to flow. The process substantially increases the volume of oil or gas from the well making it much more efficient and providing more supply for the country. Fracking is an essential part of the Oil and Gas industry and important for the country to hold a much needed domestic oil and gas supply. It helps keep prices down! An example of the lies environmental activists and zealots tell regarding oil, gas and fracking. There was a property owner in Pennsylvania who was approached by a natural gas production company who wanted to drill on her property. The royalties from the production of gas from her property would be substantial. However, after hearing all of the negative press relative to drilling and fracking she was concerned for the well being of her children. My hat is off to this lady! Rather than flatly turning down the offer she decided to do some research on the subject before making a commitment one way or the other.

She set out on a journey across the country to talk to other property owners who had oil and gas production on their land. She learned that the families living on the properties did not experience any illnesses from the wells and their water wells were not contaminated. She did find one individual who lived in an area where the water table NATURALLY contained methane. There was no drilling or production within miles of the property. The property owner actually turned on the faucet at the kitchen sink and struck a match next to the flow of water. There were flames coming from the stream of water, not like a blow torch but little puffs of flame here and there. He added a little humor to the meeting and told her his mother had one hard and fast rule that everyone honored: “Don’t smoke in the shower”. There are areas in Minnesota as well as other states that have water wells with methane contamination. Not the result of Oil and Gas production. Water from those wells can be used for a multitude purposes but are not suited for human consumption. They typically use bottled water for drinking and cooking. The well water is suitable for washing and watering the lawn as the methane will dissipate once exposed to the atmosphere.

After extensive research the property owner determined there was no threat to the health and well being of her family and entered into an agreement with the production company.

Yet another intervention in the oil and gas industry, the Biden administration attempted to outlaw the export of LNG (Liquid Natural Gas). He succeeded in halting the construction of new facilities temporarily and slowed the export for a while. This action not only impacted the industry workforce putting people out of work but also proved detrimental to our allies in Europe. Once again Natural gas is the cleanest burning fossil fuel in the world. Now with absolutely no consideration of the jobs lost, lost tax revenues and the cost of infrastructure investment made by companies exporting LNG, the government just says, sorry we're going to stop this due to Climate Change. Once again, they have never produced any valid evidence that CO2 or methane is responsible for anything related to the Climate. Most of the exported LNG goes to Europe for home heating, appliances, industry and commercial activities. In my research I've come across several articles on the subject, below are some excerpts from a couple of those articles. One from an article from Statista, a German organization that monitors such activities:

Which Country Imports the Most LNG from the U.S.?

FOSSIL FUELS

by Florian Zandt, Feb 7, 2024

"On January 26, President Joe Biden paused all approvals of new U.S. LNG export plans following increased protests from climate activists after the country became the world's top exporter of the commodity this past year. This pause will be used to conduct a review that will "look at the economic and environmental impacts of projects seeking approval to export LNG to Europe and Asia," according to reporting by Reuters. Even in the first year of Russia's invasion of Ukraine, which kicked off the scramble for other sources of natural gas, many European economies turned to the United States for their gas supply".

"France, for example, led the pack in 2022 with a share of roughly 15 percent of all exported LNG, as data from the U.S. Department of Energy shows. The United Kingdom and Spain were also "responsible for almost double-digit export shares. European countries combined received around 69 percent of all the LNG the United States exported via ship in 2022".

"The Biden administration's move to halt talks on as-of-now unapproved or future applications for the export of LNG has sparked ire among the opposition, which claimed this decision could send mixed signals to the United States allies in Europe dependent on this export flow. On February 6, U.S. lawmakers led by House Republicans held the first of two hearings in the House Energy, Climate and Grid Security Subcommittee. The second hearing is scheduled to take place on Thursday, February 8". florian.zandt@statista.com

"In 2023, the US emerged as the top exporter of LNG globally, surpassing all other nations. According to data from the US Energy Information Administration (EIA)'s Natural Gas Monthly, US LNG exports averaged 11.9 bcfd (billion cubic ft. per day) in 2023, a notable 12% increase (1.3 bcfd) compared with 2022".

Another quote is an excerpt from an article posted by the Williams Company below:

"Considering the whole value chain (LNG and natural gas supply) the cumulative economic benefits from U.S. LNG exports would range from **$1.664 trillion to $3.255 trillion** and 7.346 to 15.459 million job-years".

The above meant nothing to the Biden administration, they were focused on appeasing environmental groups that oppose the oil and gas industry. The people and the economy be damned!

Our work force here at home and our allies in Europe will suffer because of ignorant decisions of the Biden administration's leadership (or lack thereof). It is possible that you can replace the above word IGNORANT with INTENTIONAL.

Biden's administration, using the EPA, has launched an attack on the population of the U.S. using this bureaucratic agency as a weapon against the people.

Since Joe Biden was placed in office, he has single handedly attempted to destroy the U.S. Oil and Gas industry while severely increasing our dependence on foreign oil. Not just our dependence but that of Europe relative to natural gas. All of which reduces U.S income and jobs but, as mentioned, caused increased U.S trade deficits. Rather than keeping our energy dollars in house and benefiting our economy, we now have to spend them overseas to countries, for the most part, who dislike America.

His first act was to stop the Keystone pipeline construction putting about 10,000 people out of work. He then disapproved of any and all drilling for gas or oil from federal lands and waters. He then stopped the proposed development of the ANWR drill site, putting many people out of work and dramatically impacting the revenues of the State of Alaska and the income of its citizens. He has moved against natural gas utilities and (as mentioned above) appliances jeopardizing the well-being of millions of American citizens claiming gas stoves and water heaters are causing health problems for the residents. No explanation of how he arrived at that conclusion (probably scientists say). Millions upon millions of people have been raised in households using those appliances without any adverse health issues.

His next move was to issue "Methane Rules" to further hamper the oil and gas industry. His goal is obvious and his ignorance apparent. Once again if methane is a problem and causing such a "CLIMATE EMERGENCY" why are the methane gas seeps off limits to drilling.

As noted in a previous chapter - 1,900 plus climate scientists confirm there is no "CLIMATE EMERGENCY" the atmosphere has not changed content in more than a century. Carbon emissions are not causing catastrophic storms and hurricanes are not more frequent and they are not increasing in intensity. There may be more damage from storms as the increased population has demanded more housing, much of which has been built in areas that are prone to storm damage, examples would be coastal Florida, Alabama, Mississippi, Louisiana and Texas and the Eastern Seaboard. For the most part, but not exclusively, low lying coastal areas are the hardest hit when large storms make landfall. While in the mid-continent tornado "damage" has increased due to population increase in tornado prone areas. My SUV is not causing it!

Keep in mind that NONE of the doomsday predictions of the1950's, 1960's,1970's 1980's, 1990's, 2000's and the years in the 2010 to 2024 have come to pass. They have all been issued by overzealous environmentalists using their faulty computer models or just plain guess work.

A Dec. 2, 2022 article from Warner Todd Huston

(http://ipatriot.com/author warnertoddhuston/)

(https.// ipatriot.com/bidens-epa-launches-new-war-on-u-s-energy-sector-quadruples-regulatory-costs-of-carbon-emissions/) the article states "Biden administration EPA have opened a new war on the U.S energy sector by quadrupling the regulatory cost for carbon emissions".

Keep in mind that this regulation is directed at methane and CO2 emissions from oil field operations. Also keep in mind that the atmospheric content of methane is .12% and CO2 is .38% which has not changed in more than 100 years.

An excerpt from an article published by the World Bank Group "Carbon pricing is an instrument that captures the external costs of greenhouse gas (GHG) emissions—the costs of emissions that the public pays for, such as damage to crops, health care costs from heat waves and droughts, and loss of property from flooding and sea level rise—and ties them to their sources through a price, usually in the form of a price on the carbon dioxide (CO_2) emitted". HUH! No one can directly tie CO2 or methane to any change in climate. This is nothing but government over-control and out of control.

"The new estimate has sent shockwaves through the energy industry and raised the stakes for ongoing litigation being brought by Republican attorneys general in states like Louisiana that are challenging the Biden administration Social Cost of Carbon rulemaking as unconstitutional".

Per "Just the News"

(http://justthenews.com/politics-policy/environment/cha-ching-epa-quietly-quadruples-regulatory-cost-emissions-new) "The so-called Social Cost of Carbon was first set in 2009 under the then- President Barack Obama. Under Democrats it has been set for most of the last decade at $51 per metric ton. But in a rule making notice in mid-November, the EPA said it had devised a new calculation that raises that damage estimate to $190 per metric ton by 2022 standards and as much as $410 by the year 2080". This is all about money and expanding government revenues on the back of the oil industry which unknowingly filters down to the consumer. These increased fees allow them to justify hiring more agency staff to monitor industry field activities.

The figure isn't just hypothetical. Once enacted, it could affect everything from the cost of methane regulations and tailpipe emissions to the bogus "climate reparations" that President Joe Biden has committed the United States to paying the poor countries in the future.

These rules, made by bureaucrats not congress, will end up impacting every aspect of life for every U.S. citizen. You will be beaten into submission!

An article in Energy in Depth (IPAA). "The Biden White House loves California's energy solutions – shortages and high costs. The Hill opinion. "California is suffering an energy catastrophe that is almost entirely self-inflicted. Bad policies have exacerbated the impact of a dire drought, while green mania has stripped the state of the kinds of back-up solutions that are critical to shifting to renewable energy. The astonishing thing is that policymakers here in the U.S will learn nothing from California's travails. In fact, in a trip to the Golden State just days before officials warned of impending energy blackouts, Energy Secretary Jennifer Granholm gushed to an interviewer", "California is in the lead and can show the rest of the nation how it is done".(another academic achiever with no practical life experience).One wonders what part of California's "energy solutions" Granholm admires. Is it that homeowners pay 88% more for electricity than Americans elsewhere? Or that industrial users are being charged 99% more than their competitors in other states"? "Or that gasoline and diesel prices are about $2 higher than the national average"? What a model to look up to!

Biden's next move was to dismantle the U.S trucking industry. A move that will completely destroy the economy of the U.S. As stated in the previous Chapter, "nothing moves without a truck". The technology to convert all diesel trucks to Electric Vehicles is currently extremely costly and there is no end to the high cost in sight. Putting the infrastructure necessary to bring the cost down will take decades if it is feasible at all.

The proposed "RULES" once again, from the EPA not congress, are impossible to attain and will put countless citizens out of work. The supply chain will be no more. Plant you garden and raise some livestock if you feel you will need to eat in the future. Someone may revive the wagon making industry. Moving from the east to west or visa-versa will be much easier via wagon now than it was in the 1800"s as we have interstate highways that will be vacant. No more dependence on the Oregon Trail, probably easier on wagon wheels as well. Also, probably won't experience the events of the Donner party. However, the EPA will likely put "RULES" in place dealing with manure removal and disposal and impose regulations due to methane emissions from the team of horses/oxen pulling the wagons.

As if California gas prices were not high enough the city of Richmond CA introduced a ballot initiative to tax the Chevron Richmond Refinery at a rate of $1 per barrel of crude oil throughput. The Chevron plant is one of the largest refineries in the state at 245,000 barrels of oil per day. This translates to just under $90 million in new taxes per year for the refinery. If successful, you can bet all of the other communities where refineries are located will follow suit. The refinery cannot absorb the additional tax burden therefore Chevron will have to increase the price of their wholesale products much of which is sold locally in Richmond and surrounding communities. **Update:** Chevron has paid the City of Richmond over a half million dollars to keep the measure off the ballot.

Can anyone say EXTORTION! Will this set a precedent for other communities having refineries?

Incidentally, Chevron has moved its headquarters out of California after 140 years to Texas.

Every governmental agency has to get involved. In Texas, the Fish and Wildlife Service issued a rule placing the Dunes Sagebrush Lizard in the Endangered Species Act. This rule will impact exploration activities in the Permian Basin, one of the largest and most prolific oil producing areas in the country. This rule could be devastating to the Texas economy and the country due to lost jobs, revenues from the whole chain of workers from drilling companies, field workers, laboratory workers, trucking/ pipeline workers as well as the U.S. energy security. These government agency personnel just sit around looking for things they can interject in order to disrupt the oil industry. The Permian Basin has been producing oil and gas for more than a century, never having heard of any issues with Dunes Sagebrush Lizards.

I would guess that as many people are familiar with the Dunes Sagebrush Lizard as are familiar with the California Blunt-Nosed Leopard Lizard. The company I worked for, in the early 1990s, was involved in laying a short 200' pipeline in the lower San Joaquin Valley, California, in the middle of an oil field. The purpose was to connect an oil gathering tank farm into a common carrier pipeline with the intent of reducing truck traffic from the site. This pipeline would reduce truck traffic and provide a more economical means of getting the oil to market. We were informed, by local agencies, that we had to have an environmental impact study done and as part of that study a "scientist " would have to walk the proposed right-of-way for the pipeline to ensure there were no endangered species living on the right-of-way. Keep in mind we were in the middle of an oil field with hundreds of oil producing wells surrounding the proposed right-of-way. We contacted this so-called "scientist" and had him do his inspection. He walked the right-of-way all 200 feet of it and told us that he found some Blunt-Nosed Leopard Lizard droppings on the right-of-way. How he could tell the difference between that and a grain of sand is beyond me. He then informed us that he would have to relocate the lizard. He was asked if he was aware of what we were going to do, which was to dig a 2' wide, 4' deep and 200' long trench with a backhoe. Lay an 8" welded pipe in the trench then back-fill the trench, at which point it was mentioned that if there was a lizard on the right-of-way, he would likely relocate himself as soon as the digging started. Logic did not prevail!

As a result, we had to pay the "scientist" $10,000 to "relocate" the lizard. He came out to the site in the evening with his trap. There was no one on site to witness that. The next morning when our company personnel arrived at the site they found our "scientist" coming toward the tank farm, trap cage in hand declaring that he caught the lizard. My guess is that it was his pet lizard that he kept in the cage at home and used it to swindle others with the same type of project in mind. These rules and regulations put forth by bureaucratic agencies provide opportunity for a great deal of corruption.

California claims they will be carbon neutral by 2035. First, that is a total impossibility as discussed earlier but how many California citizens are going to be willing to give up beer and CARBON-ATED soft drinks. Yes, beer and soft drinks have CO2 in them. Every time you open a beer or Coke/Pepsi type soft drink that "fssst" you hear is CO2 escaping the container into the atmosphere. Is it beginning to soak in how ridiculous these proposals are??

Below are some quotes from an article at www.libertybugle.com "Americans-likely-to be-tracked-for- co2-emissions-under-secs-new-climate-rule-Consumer Research" the article talks about a new proposed rule by none other than the environmentally sensitive and responsible Securities and Exchange Commission (SEC)??? **"The Enhancement of Standardization of Climate Related Disclosures".** What the hell is the SEC doing in the climate business???"The nearly 500 page rule would require SEC registrants, mostly public companies, investment advisers and broker dealers to report certain climate related information including greenhouse gas (GHG) emissions". Other than CO2 and Methane, what are greenhouse gases? As noted in Chapter 3 there is only .38% of CO2 and .12% of methane in the atmosphere, the 1.95% balance of the blanket surrounding the earth is WATER VAPOR.

"The GHG emissions are categorized into three scopes. "Scope -1- is registrants direct GHG emissions". (I'm assuming this is the emissions of the manufacturing process). "Scope - 2 - is its indirect GHG emissions from the purchase of electricity and other forms of energy". (It appears the manufacturer is going to be responsible for the emissions of the power plants that supply their electricity) Scope – 3- is indirect emissions from upstream and downstream in a registrant's value chain". HUH!??

Scope three requires these companies estimate the carbon output of the use of their product by the consumer, which means they're going to have to go out into the field to talk to "consumers". At what cost and the SEC is going to be tasked to monitor this activity? "WILL HILD, executive director of Consumers' Research, America's oldest consumer protection organization, said in an interview with NTD"s (news organization) "Fresh Look America" program on July 12".

"Let's say you bought an internal combustion engine lawn mower. The lawn mower company will need to know how many times you mow your lawn. They are going to have to go out and ask people and research that, how long does it take to mow your lawn? What Is the acreage? So, you can see how this starts to lay the groundwork for scoring actual activities of individual people". Hild said. "Can you imagine the size of the bureaucracy necessary to implement this proposal? The cost to publicly traded companies will be astronomical. That added cost will increase the cost of their products and reduce return on investment for those willing to invest. In short, it will cripple every industry in the country".

Beyond the corporate impact every single community in the United States will have its own branch of bureaucrats to monitor your personal activities. Can anyone spell GESTAPO? God help you if you mow your law more than your initial disclosure.

This is big brother on steroids. It is the start of anything and everything you do will be monitored and evaluated by the government. Your personal freedoms will evaporate. Can you imagine what this kind of operation will cost the manufacturer of goods, the government agency responsible for oversight (tax $'s through the roof) and the consumer? The cost of goods will skyrocket and the taxes to pay for the necessary government employees will be exorbitant. The size of the government will grow exponentially.

Know this, the whole "Global Warming" movement is all about control and has nothing to do with the climate. The enemies of the U.S. have signed on to this idea in order to propagate fear into the minds of the uneducated in an effort to curb the success of the U.S and to bring us down. Many of our elected officials are complicit in the movement.

Any current or future change in the climate are natural occurrences and there is nothing the government is going to do about it regardless of how many asinine "RULES" they make.

There is an article in the publication "Republican Investor", That states "The United Nations Climate Change Agenda is Designed to Starve and Kill You Off, Not Save The Planet". The article speaks to all of the international rules and regulations focused on the environmental impact of farming. The end result is a reduction in food supplies. "I would remind the readers of this that we are a "sovereign nation" and should not be dictated to by the U.N".

You will note that while these international organizations insist that the United States conform to rules and regulations they propose, they will not comply with what they want the U.S to do. This is to add cost to our industries making our products more expensive to where we are unable to compete in the world market. This opens all markets up to foreign countries who will not be complying with the restrictions placed on U.S manufactures. This would be extremely detrimental to the U.S. trade deficit. All of our cash for goods would be exported purchasing goods from around the world while our corporations go bankrupt. There would be no job market. We are a sovereign nation and our politicians should not try to hold us to rules and edicts from foreign organizations.

AND SO, IT BEGINS -The latest from California is the California Energy Commission is proposing that they take over all oil refining assets in the state of California. This is a Russian /Hitler move. Once in control they will reduce the available motor fuels to the public in an effort to force people into electric cars, even though they don't have the grid capacity to sustain the demand for the necessary power. You will likely see gasoline prices go to $20 per gallon, assuming you can find a station. They will also phase out the oil producing industry in the state. There, currently, is a little less than 300,000 barrels per day of oil production in California, much of the production is owned by small independent operators. The only home for that production is California refineries. There are no pipelines to move crude out of the state and trucking is a non-starter. Once controlled by the government they will have the opportunity to refuse acceptance of state production. They will shut-in every lease in the state. Can you say COMMUNISM, put that along-side of STATE BANKRUPTCY. Absolutely no regard for the tax revenues generated by the Oil and Gas industry and the people that industry employs.

Below is an excerpt from the publication "Energy in Depth" relative to the above California Energy Commission's proposal:

California Energy Commission proposes state takeover of oil refineries. California Globe. Just as Chevron Oil company announced that it is moving its headquarters to Houston Texas from San Ramon California, California Energy Commission regulators announced proposed government controls of the petroleum industry, ostensibly in order to combat future energy price surges, according to a report released August 1, 2024 by the CEC. Despite California's radical and very accelerated green agenda which does not include oil and gas as fuels, the CEC fully expects some of California's nine oil refineries to be shuttered due to falling demand, which would give the remaining refineries increased pricing power and raise the possibility of a surge in gas prices, the study said.

This proposal does not only impact the oil industry in California. It impacts every single industry that operates in the state, especially agriculture. Agriculture production from California does not only provide food for Californians but for many states across the nation. You will see major increases in produce prices in the grocery store once they start restricting access to motor fuels. Not only will production drop off but the ability to transport the produce to market will be affected.

The scary part is that the liberal politicians in Washington D.C use California as an environmental model for the country. Will we soon become Venezuela?? If so, where will we go? Our borders are wide open but no other country will allow us to just come on in.

More from the Biden administration as they pushed to electrify everything but the kitchen sink. We now have to have electric stoves, water heaters, heating and cooling systems, cars, lawn mowers, etc. With all those changes we become more dependent on China. While at the same time we have to get rid of all of our natural gas appliances. Pry tell is there a recycle bin for those appliances. Aside from the fact that natural gas appliances are far more efficient than electric appliances. Why do all restaurants cook with gas stoves? They are easier to control heat and they are more efficient.

In a report by "Power the Future" they address Biden's green agenda and how it is making us more dependent on "China's growing dominance".

Specific items the article mentions are:

"The bipartisan resolution of disapproval" passed by the Democratic Senate in May 2023 targeting Biden's decision to allow tariff waivers for solar modules and cells from Cambodia, Malaysia, Vietnam, and Thailand that gave China a free pass". This has a definite impact on jobs in the U.S.

"The Consumer Product Safety Commission (CPSC) exploring potential bans on gas stoves; The Environmental Protection Agency(EPA) proposing the de facto EV mandate, one that would push EV's from nearly 6% today to nearly 70% by 2032; the $369 billion dollars' worth of green handouts included in Biden's Inflation Reduction Act (IRA which should have been entitled "The Green New

Deal"); "circumventing West Virginia Senator Joe Manchen's requirements included in the IRA; Biden's Department of Energy conditionally selecting Microvast, a battery company with ties to the Communist Party of China, to receive a $200 million Grant; The Biden Administration's consistent rejection of permits for significant new mining projects, despite America's over-reliance on foreign sources and adversarial nations for critical minerals and materials; Along with his son Hunter, Biden's active support of Chinese mining interest abroad" China and the Taliban cut a deal for China to access the massive amount of mineral deposits in Afghanistan".(as mentioned in a previous chapter regarding the Afghanistan withdrawal).

All of the above subjects can be found in much more detail at https://powerthefuture/where-green-meets-red-how-the environmental-agenda-is-making-America-dependent-on-china.

Biden declared a "CLIMATE EMERGENCY" to give him an excuse to limit oil and gas production and export. Can anyone tell me how you identify a "Climate Emergency". What has to happen short of a hurricane or tornado to be classified as a "Climate Emergency"? Hurricanes and tornados I would consider being a climate emergency though they are short lived they do have a long term effect. Hamstring or eliminating the oil and natural gas industry is not going to stop hurricanes or tornados from happening, it will only destroy the economy. Is there any other condition that can be called a "Climate Emergency"? Perhaps wildfires but like hurricanes and tornadoes those aren't caused by my SUV, for the most part they are caused by arsonists.

The Biden Administration initiated an all - electric mandate to phase out internal combustion engine vehicles (ICEV). Donald Trump's response to that is that there will be a "Bloodbath" in the auto industry. In support of Trump's theory Ford Motor Company announced that their Ev division lost $4.7 billion in 2023 and that in the first quarter 2024 they were losing $132,000 per unit sold. Needless to say, that cannot be sustainable. China will be the primary electric car manufacturer. General Motors lost $3.3 billion in 2022 but did much better in 2023 only losing $2.5 billion.

As a footnote to Biden's Chinese connection Warner Todd Huston points out companies that were once U.S. businesses now owned by China: "Dirt Devil Vacuums, Smithfield Hams, Forbes Magazine, GENERAL MOTORS, GENERAL ELECTRIC, Motorola, IBM, TESLA, Instagram Micro, OmniVision Technologies inc., Segway Inc, Waldorf Astoria Hotels, Legendary Entertainment Group, Hoover Vacuums, Fidelity & Guaranty Life, Brookstone Inc, Dairy Farms of America Inc".

An article in IPAA's Energy in Depth shows how Biden's war on the Oil and Gas industry harms the U.S. economy relative to liquid natural gas exports. These sales bring in billions of dollars to American companies annually. Also, it disrupts the energy supply of our allies.

Joe Biden's pyrrhic fight against LNG exports harms the U.S. economy. Real Clear Market. The Biden Administration's January 2024 pause on liquified natural gas (LNG) exports to non-free trade agreement countries merely formalized a de facto ban that had been in place since the beginning of Biden's term. This policy is at odds with our official foreign policy goal of helping our European allies wean themselves from Russian energy and has hindered the creation of thousands of high-paying jobs. Ending this policy should be a high priority for the new Trump Administration. FERC and DOE share responsibility for approving LNG projects. FERC has the responsibility to approve the location and construction of the project and the DOE's legal responsibility is to determine whether anyone opposing exports has adequately proven that the proposed exports are "inconsistent with the public interest."

In a meeting with Speaker of the House Mike Johnson, Biden denied signing the executive order that put LNG facilities on hold ceasing LNG exports. He did not remember signing it three weeks prior to the meeting. There is a chance he didn't sign it, rather a staff member intervened.

This chapter could go on for thousands of pages pointing out how our own government is selling us out to foreign entities while making our industries conform to rules and regulations set by foreign countries designed to destroy our businesses in the name of CLIMATE CHANGE.

Chapter 8
Government Waste and the Attack on the Automobile and Oil and Gas Industries

As outlined above in a previous chapter the Biden administration in, 2021 provided $7.5 Billion in federal tax dollar grants to build charging stations across the country. To date there have been 8 stations built with that grant money and none so far are known to be operational.

According to some internet information, currently there are about 168,000 charging stations in the U.S. Some sites say 63,000 some say 180,000. It is difficult to say exactly how many there are and which are fast chargers, slow chargers and which are not operational. This is problematic for travelers using EV's.

Quoting the Article – "The Cold Truth Initiative" by Tom Luongo: "As mentioned earlier the Solondra deal concocted by then President Obama, has more to the story than Solondra going bankrupt. The $500+ million ended up going to Goldman Sachs and George Kaiser. Both of those firms contributed heavily to the Obama reelection campaign to the tune of $1.25 million".

"Additional taxpayer waste occurred in 2010 wherein another $400 million went to Abound Solar. That resulted in bankruptcy as well. But investors in Abound seemed to do just fine. Investors like billionaire heiress Patricia Stryker. Stryker contributed $500,000 to the Coalition for Progress while throwing $85,000 toward former President Obama's inaugural Committee. It is a mere coincidence that the government handed a company she invested in $400 million just before bankruptcy."

"There's also A123 Systems who paid one lobbying firm $970,000 to secure money from the government. They then received $279 million in federal assistance. The CEO of A123 Systems went on to fund multiple Democratic senators and contributed to former president Obama's campaign".

"First Solar received $646 million in government loan guarantees and has since contributed $180,000 to democrat campaigns".

" GE is notorious for spending tens of millions of dollars a year to "buy" green energy credits for its wind turbines and other green technologies – credits which helped them pay ZERO taxes in 2011". By the way GE is owned by the Chinese multinational home appliances company Haier.

The Daily Caller News Foundation published an article:

(https://www.whitehouse.gov/briefing-room/statements-releases/2023/04/17 fact-sheet-biden-harris-administratio-annouonces-new-private-and-public-sector-investments-for-affordable-electric-vehicles/)

"As part of its goal of having EVs make up at least 50% of all new car sales by 2030, the Biden- Harris administration has lavished (https://dailycalle.com/2023/08/31/biden billions-retrofit-auto-plants-electric-vehicles/) subsidies on automakers, making $12 BILLIONS in taxpayer funds available for manufacturers to retrofit their plants in August of 2023 in what the white house claimed would create over 2,900 new high quality jobs". "GM claimed a retrofitting process (https://www.inc.com/reuters/gm-plans-to-lay-off-1700-workers-ar-kansas-plant.htm) – "the installation of new tooling"-was the reason for the Fairfax, Kansas, temporary layoffs".

"It may sound like hyperbole, but it's not an exaggeration to say that we risk the wholesale destruction of the U.S. auto industry," Mark Mills, Executive Director of the National Center for Energy Analysis, told DCNF (Daily Caller News Foundation) "It is obvious that the subsidies are neither accelerating any putative, significant cost declines in EVs, nor making it feasible to completely restructure the auto industry. The subsidies, instead, are increasing the dependence on foreign materials and creating foreign jobs especially in China."

While the taxpayers' foot the bill for subsidies to automakers in order to manufacture vehicles that the general public is not interested in purchasing. We hear of the automakers losing tens of thousands of dollars per vehicle. The inventories of EVs are increasing across the country as manufacturers can't seem to find buyers. Further there are stories relative to EV owners having buyer's remorse and wanting to return to internal combustion engine vehicles or hybrids. The problem is that there is no used EV market, the value for an EV once it initially leaves the dealers lot drops like a rock.

There has been a rush to buy EV's by uninformed buyers who put their trust in the government that is telling them that EVs are environmentally friendly and if they buy one, they are saving the planet. As demonstrated in the previous chapters, nothing could be further from the truth.

The $22 billion spent on Climate change every year ($42,000 per day) as listed in a previous chapter is just the tip of the iceberg. The real cost after considering all of the subsidies, government rules and regulations, imposed on various industries, associated with so-called climate change moves the cost considerably higher. According to FORBES Magazine, "the true cost is up to $1.75 trillion per year ($3,329,528 per minute) of your tax dollars going to waste. Not only are your tax dollars going to these companies for a non issue but an additional burden is placed on the taxpayers in the form of cost of goods".

"The reality is, these initiatives have a ripple effect. Mainly the regulations from those agencies (government bureaucratic agencies such as the Environmental Protection Agency) that shackle free enterprise and force us to rely on foreign energy".

We are now looking at spending $50 trillion over the next 25 years and the Federal government cannot identify where that money will be spent and on what. It's just the never ending battle against global warming, global cooling(?) and Climate change. The earths' atmosphere has not changed composition in more than a century, so what is causing the change in the climate that requires such massive spending. How do they intend on "correcting the climate"??? Especially if the politicians don't know what is causing their perceived "climate emergency". True climate scientists are well aware of what causes heat waves in the summer and very cold spells in the winter in some areas. In fact, they can actually predict weather anomalies based on “Solar Cycles”. There is nothing the federal government is going to do about it no matter how much money they throw down the drain. The slight warming the earth is experiencing currently is also happening to the other planets in the solar system, it is doubtful that my SUV is causing it. The one common denominator is the SUN.

We in the United States have the ability to be energy independent. During the last administration (Trump 45) when oil production was encouraged, the country was energy independent with the exception of California which remained 75+% dependent on foreign oil. California has about 1.6 million barrels per day of crude oil refining capacity (some refineries are now converting to Biofuels). The state produced about 307,000 barrels per day (bpd) of crude oil in 2023 that number has now dropped to about 295,000 barrels per day in 2024 (info from the U.S. Energy Information Administration EIA). California has one of the largest proven recoverable oil and gas reserves in the country, in several different formations. However, local politics make it impossible to explore and drill. California politicians want oil out of the state. They have no idea of the impact on the state residents nor do they care. California is one of the highest agriculture producing states in the country (fruits, nuts and vegetables) heavy on the fruits and nuts, in and out of the fields. It is said that if the San Joaquin Valley were to be planted with various crops from the southern end of the valley (Grapevine) north to Sacramento, and the Sierra Nevada mountains west to the Coastal Range, the produce could feed the nation. Lack of water prevents that from happening and soon the lack of petroleum will eliminate existing farms. Converting farm equipment to battery powered vehicles coupled with distribution equipment being electric will make the cost of produce skyrocket to unaffordable prices. Want to pay $6 for a tomato???

As mentioned in an earlier chapter, one municipality, Richmond CA, was trying to get an initiative on the ballot to tax a local refinery, owned by Chevron at the rate of $1 per barrel of crude oil throughput. The Chevron Richmond refinery is a 245,000 barrel per day facility (about $90 million per year in new taxes). One of the largest refineries in the state. The residents now complain about the high price of gasoline and diesel in California, a tax such as what was proposed would be passed on directly to the consumer. Had it passed it would have been the camel's nose under the tent. Soon other municipalities would follow suit and the cost of gasoline would go through the roof. As mentioned in a previous chapter the city of Richmond extorted $500,000 from Chevron to keep the initiative off the ballot. One can only hope that will not set a precedent encouraging other municipalities to attempt the same extortion.

Drilling the ANWR in Alaska would produce about 1 million barrels per day which would be delivered to the port of Valdez by way of the Trans Alaska Pipeline from the North Slope. Most of that oil would be delivered to refineries in California, a major win for the country as we would nearly be energy independent coast to coast. The ANWR reserve has been placed off limits by the Biden administration making U.S.

refiners buy foreign oil. More cash to foreign countries increasing the trade deficit, more expense to U.S. citizens.

Once again Energy In Depth identifies the issue:

Alaska sues Biden administration over oil and gas leases in Arctic refuge. Reuters. The U.S. state of Alaska has sued the Biden administration for what it calls violations of a Congressional directive to allow oil and gas development in a portion of the federal Arctic National Wildlife Refuge (ANWR). Monday's lawsuit in the U.S. District Court in Alaska challenges the federal government's December 2024 decision to offer oil and gas drilling leases in an area known as the coastal plain with restrictions.

We are now facing, through government intervention, as much as $50 Trillion tax payer funds that we don't have, to be spent on so-called climate change and no one can identify the benefits of the expenditure. This according to Senator John Kennedy's congressional interview of (Biden's) Deputy Secretary of the Department of Energy, as outlined in a previous chapter. The benefit to worldwide temperatures cannot be measured because it is non-existent. But the U.S. Department of Energy and the many braindead politicians believe we should spend the money anyway. We need to conquer the non-existent "Climate Emergency".

It is not uncommon for politicians to use assets belonging to the people for their own political gain. Case in point, Joe Biden reduced the volume of our strategic petroleum reserves in an effort to lower gas prices for political points. In the process he extracted several million barrels of oil from the petroleum reserve, and sold 500,000 barrels of that oil to our enemies, the Chinese. The U.S Strategic petroleum reserve is oil set aside for national emergencies such as wars, to enable our military to continue to have fuel in the event of supply disruptions. Selling the reserves puts the country at risk. The reserves are not in place to sell into the market to reduce high fuel prices caused by the administration. Biden's policies are what caused the high fuel prices at the pump with his anti-oil position. Those policies have made us much more dependent on foreign oil. The administration then looked to sell additional oil from the reserve to lower prices in time for the 2024 presidential election. Once again for political gain.

In addition, Biden drew down the Northeast gasoline reserves which are stored to protect the country in times of emergencies. Instead, he decided to reduce the reserves in an attempt to lower gas prices at the pump before the election. One again for political gain, an all about "me" (Biden) agenda. The people of the country do not matter. Our nation's energy and security should not be used as a political football.

The U.S government is in the process of trying to push electric vehicle ownership from its current level of 6% of U.S. vehicles to 70% by the year 2032, just about 7 years from now. Not only is there insufficient electrical infrastructure to accomplish that goal. What they have not considered is that the tire wear on EV's is substantially higher than on gas powered vehicles. This is due to the excessive weight of the EV. Around 3,000 pounds more than the gas powered passenger vehicle. The tire wear creates much more emissions than that of gas powered vehicle exhaust. An organization by the name of Emissions Analytics states "losing tread pollution from tire wear of EV's is now 1,850 times worse than exhaust emissions" At 6% EV's on the road this may be manageable but at 70% it would be overwhelming. This also flies in the face of those who claim EV's are more environmentally friendly, especially when considering all of the mineral excavation and processing necessary to produce one battery.

In addition to all the negatives listed above relative to EV's and "Climate Change" we are now seeing cities, counties and states entering into lawsuits against oil and gas producers for excessive amounts of money. According to an article in the Energy in Depth publication (Wed. July 17) where they quote E&E News headline "Puerto Rico files a $1 billion climate lawsuit against oil companies". Puerto Rico is joining the ranks of more than two dozen cities, counties and states that are seeking compensation from oil and gas companies for the effects of climate change. Filed Monday in the Court of First Instance of San Juan, the lawsuit claims that despite knowing their products would pollute the island and cause temperature to warm, Exxon, Chevron, and other companies marketed and promoted oil and gas using unfair and deceptive trade practices". We have seen these claims before, it is about money for the entity suing. It has nothing to do with climate change. There is no proof that oil and gas is causing their perceived problems. How could they possibly quantify the impact? Could you imagine the lawsuit if those oil companies decide to discontinue product sales to those communities that are suing?

Putting the onus on oil companies for wildfires and or natural disasters is a gross abuse of political power. The legislators in some states as well as Puerto Rico are claiming that oil companies are responsible for wildfire and natural disasters while not presenting one iota of evidence to support the claim. Now they want to be able to sue the companies and have them pay to repair the damages caused by nature. Many fires are caused by lightning and utility power lines not to mention arsonists but let's make Oil and Gas companies pay for the damage, they have plenty of money! The articles below are from the IPAA publication Energy In Depth.

"Big oil pushed to kill bill that would have made them pay for wildfire disasters. The Guardian. In the year preceding the devastating Los Angeles county wildfires, big oil fiercely lobbied to kill a "polluter pay" bill that moved through the California senate and would have forced major fossil fuel companies to help cover the costs of climate disasters. Fossil-fuel industry lobbying in California spiked to record levels during the 2023-24 legislative session, and the polluter pay bill was among the most targeted pieces of legislation, a Guardian review of state lobby filings found"

"L.A. fires revive calls for a 'Climate Superfund' law in California. NYT. Vermont and New York recently became the first states to create "Climate Superfund" laws, which will require energy companies to help pay for the costs of dealing with extreme weather and climate change. As wildfires devastate the Los Angeles area, some are asking if California could become the third".

Oil and gas companies should pay for the Los Angeles fires. New York Times. Opinion. Major oil and gas companies have known for decades that burning their products could lead to "potentially catastrophic events," like the higher temperatures and abnormally dry conditions that fed the fires still being battled in Los Angeles. Exxon scientists warned about this internally starting in the 1970s. Instead of disclosing these risks, big oil companies deliberately misled the public, policymakers.

California bill would let insurers sue oil companies to avoid raising rates. Politico. California lawmakers introduced a first-in-the-nation bill Monday to create a clear legal pathway for individuals and insurance companies to sue oil companies for damages after fires like the ones that destroyed whole neighborhoods in Los Angeles this month. What happened: State Sens. Scott Wiener and Sasha Renée Pérez introduced SB 222, the Affordable Insurance and Climate Recovery Act, on Monday with the goal of shifting the burden of increasing insurance costs from California ratepayers to fossil fuel companies.

Once again, out of control and absolutely ignorant California legislators are trying to force a bill that holds Oil and Gas companies responsible for fires in the state. I would counter the proposed bill with holding the legislators responsible in that for more than 50 years they have voted down proposals for building additional reservoirs to capture snow melt runoff which would make water available during drought conditions and to have water resources available for fire departments to extinguish fires.

Can someone kindly tell me how the use of petroleum products by the general public can cause "CATASTROPHIC EVENTS"??? Except for an arsonist using gasoline as an accelerant.

Just how would they and who would they propose to sue? Which oil company is responsible for which fire? Did ExxonMobil cause the LA fires or was it, Chevron? I was always told that in order to sue someone you need some proof of culpability. How exactly will they frame the blame? And what evidence will be presented?

Once again, we see deep pockets witch hunt. The politicians let the people down by being unprepared and allowing conditions to get out of control and insurance companies are left holding the bag. Then the insurance companies can't foot the bill due to the magnitude of the events and will go bankrupt if forced to provide the funds to rebuild. Therefore, we must find someone with money, matters not if they were involved, they just have to have a lot of money. This would set an extremely dangerous precedent. What industry would be next? Do we target Car dealerships? How about wine makers? How about breweries? Oh yeah, Doctors! They all have money! Where would the money go once the state collected? My guess it would not go to the victims of the fires and politicians would find a way to waste it.

Not only are Oil and Gas companies blamed for Wildfires but they are now the target for all perceived climate conditions. And are expected to foot the bill for rising tides and storm frequency. Below another excerpt from Energy In Depth wherein the Supreme Court supports frivolous lawsuits against oil companies:

Supreme Court clears path for climate lawsuits against Big Oil. E&E News. The Supreme Court has given a fresh boost to state, county and city officials looking to hold the oil and gas industry financially accountable for its contributions to climate change. In an order issued Monday, the justices declined to take up the fossil fuel industry's claims that lawsuits asking companies to help communities foot the bill for rising tides and worsening storms pose a threat to a sector that is "vital to economic and national security." The high court's denial — with no justices signaling a dissent — is "bad news" for the oil industry, which currently faces 28 active climate liability lawsuits, some of which are likely to go to trial, said Michael Gerrard, faculty director of Columbia University's Sabin Center for Climate Change Law. Also covered by Associated Press and Los Angeles Times.

So, oil companies are now the catch all for anything that impacts anyone relative to the environment. All of these lawsuits cost money! A suit filed by a government agency is on the taxpayer dime for the government legal expenses but it also cost the oil company millions to defend themselves. Those defense funds come from the customer base in the form of increased product prices. These Lawsuits are a double edged sword with the citizens being the victims. Likewise, a suit filed by a private entity against the government has the same impact on taxpayers and consumers. Loser pays legislation would likely stop this insanity. Another article from Energy In Depth. This one makes mention of the proposed legislation mentioned above but it goes on to question the common sense of the proposal relative to some of the benefits of oil and gas.

Democrats In California attack energy producers who make firefighting possible. American Energy Alliance. Two Democratic lawmakers introduced legislation in California to allow residents who have damages from natural disasters to sue oil and gas companies. California would be the first U.S. state to allow such lawsuits if approved. Of course, there is no mention of the benefits of fossil fuels, including supplying life-saving heat in winter, air conditioning in summer, food on the table, and economic prosperity from supporting industrial activity. Oil companies have already survived several climate lawsuits claiming that they have deceived the public regarding the risks of fossil fuels.

Energy In Depth points out that California wackos are now suing the Bureau of Land Management for issuing drilling permits on Federal lands in the state. It should be noted that federal lands in California are largely UNPOPULATED and mostly desert.

Interior faces challenge to California oil, gas drilling permits. Bloomberg Law. Environmental and health organizations are challenging the Interior Department's continued approval of drilling permits for new oil and gas wells on public land in California's San Joaquin Valley, which already faces pollution from other sources. Interior's Bureau of Land Management made the approvals without accounting for the cumulative effects of expanding oil and gas drilling beyond existing wells, the Sierra Club, the Central California Asthma Collaborative, and other groups said in their complaint. The agency also failed to seek meaningful input from the communities most impacted by its permitting decisions, the groups claimed in their suit filed.

Environmentalists sue after feds allow new drilling permits in one of the country's most polluted areas. Independent. A slew of health and environmental groups are suing the Bureau of Land Management over its approval of permits for new oil and gas drilling on Central California's public lands - which is home to some of the country's most polluted cities. A complaint filed in the U.S. Eastern District Court of California claims that the federal agency's decision is at the expense of public health, the environment and the law. It adds that the bureau had never analyzed the harms of its well approvals on nearby communities.

The second article speaks of "some of the countries most polluted cities". News flash the BLM lands in the San Joaquin Valley are sparsely populated so exactly what polluted cities are they referring to? Right now, the most polluted cities in California are Los Angeles and San Francisco due to the homeless population (human waste pollution).

California is not only a state with whacko legislators. Once again Energy in Depth identifies New York state government working against the people:

Hochul's fracking foolishness locks $1 trillion of energy underground — and hurts New Yorkers. New York Post. Opinion. Hydraulic fracturing, better known as fracking, has transformed the United States into the Saudi Arabia of natural gas — but not in New York, where foolish politicians like Gov. Hochul stubbornly refuse to tap the enormous wealth beneath our feet. Two giant shale formations sit beneath much of northeastern North America.

The uneducated are put in a position to make laws and rules that impact the masses. The above from the state of New York prevents the citizens of the state from having low cost energy for home heating. Natural gas is far cheaper and far cleaner than fuel oil which is what is primarily used to heat homes in upstate New York. They are quite literally sitting on trillions of cubic feet of natural gas and the government will not allow energy companies to drill.

What needs to be understood here is that Major Oil Companies are large corporations having a multitude of stockholders who the company management answer to. Large corporations collect the taxes they pay from their customer base. If the government imposes new financial exposures and these corporations are forced to pay for natural disasters which they are in no way responsible for you can count on the products they sell will come with a much higher price to the consumer. The price at the pump as well as all things petroleum based could easily double (take a look back to Chapter one for petroleum based products). But of course, that would be big greedy oil company price gouging. Nothing to do with government intervention. What happens to the source of funds when California goes "all electric and carbon free" (which is impossible). Who pays for the fires then???

They use the term "Wildfires" as an all-encompassing category and blame them on Oil and Gas companies. Those companies should therefore pay for the restoration of damage done. There is no mention of the fires started by arsonists or lightning strikes. There is no proof at all that these companies are in any way at fault, natural disasters are just that NATURAL and those caused by human intervention such as power company equipment failures or arsonists are certainly not the responsibility of oil companies. What of the fuel supplied by oil companies for the fire trucks and aircraft that extinguish the fires? Without that equipment the fires would be left to nature resulting in far more damage.

When governments get involved with industries that they know nothing about, the only possible outcome is a catastrophe which is generally extremely costly to the taxpayer. A couple of articles found in Energy In Depth demonstrate that below:

Gas prices and water fights: California environment 2024 year in review. SMDP. California strived to maintain its environmental leadership in 2024, with voters approving a $10 billion environmental bond and the state advancing its renewable energy goals — all while navigating a major budget deficit. The Legislature approved only a few new major environmental policies as the deficit left environmental groups and renewable energy advocates clamoring for increased spending.

The $10 billion bond must be paid back to the bond purchasers with a set interest rate. So just how much is it really costing the taxpayer.

Gov. Gavin Newsom and the Legislature scaled back their "California Climate Commitment" spending package to $44.6 billion from $54.3 billion.

How generous! I'll bet that move went over big with the climate wackos!

Here's the real hockey stick. Robert Bryce Substack. The Inflation Reduction Act, which became law in 2022 thanks to a single vote cast by Kamala Harris, has turned into a run on the Treasury. New Treasury Department numbers show that soaring federal handouts for wind & solar dwarf all other energy-related provisions in the tax code and will cost taxpayers $421 billion by 2034.

More cost exposure to the taxpayer, as mentioned above, in an effort to disguise his war against the oil industry, Biden proposed to draw down the national gasoline reserve in an attempt to lower gas prices just before the election. According to the publication "Energy in Depth" who quote POLITICO In an article published in January 2023 "Republicans are aiming to neutralize one of the main tools that President Joe Biden used to lower gasoline prices before last years' elections- his prolific releases of oil from the nation's Strategic Petroleum Reserve. The house GOP is calling a vote this week on legislation, H.R. 21, that would prohibit releases from the underground petroleum stockpile unless the government approves a corresponding increase in domestic gas and oil production on federal lands. Two weeks before the House passed legislation that would ban sales from the reserve to China". (Don't know why that would require legislation, I was always under the impression that aiding and abetting our enemies was treason).

An, article published by American Action News

https//americanactionnews.com/featurd/2024/08/12/countys-electric-bus-ptogram-resulted-in-millions-of-dollars-in-wasteful-spending.

“An investigation into a Maryland school system’s electric bus initiative found that it resulted in millions of dollars in “wasteful spending” due to issues with the vehicles, according to a report from the county’s Office of Inspector General(OIG) released in late July”.

“In February 2021 Montgomery County Public Schools (MCPS) in Maryland announced it was replacing 326 diesel school buses with electric buses as part of its sustainability initiative, the county OIG concluded in a report. Due to delayed deliveries and frequent mechanical issues with the electric buses, the county has had to partially resort back to diesel buses despite the roughly $170 million it paid for the new fleet”. This equals about $520,000 per bus.

“All of the buses received during fiscal years (FY) 2022-2024 were delivered beyond the contractually required delivery date, the OIG report stated. As a result, MCPS spent over $14 million to acquire diesel buses to compensate for not receiving the anticipated electric school buses””

“In addition to delays many of the electric buses have been “inoperable for extended periods” with the OIG estimating that “MCPS should have assessed the (electric bus) contractor approximately $372,000 for failing to provide serviceable buses between FY 2022 and 2024”.

“MCPS financial manual defines waste as “the extravagant, careless, or needless expenditure of MCPS funds, or the consumption of MCPS resources that result from deficient practices, systems, controls, or decisions”. The report reads “MCPS”s failure to hold the contractor accountable to the terms of the contract and their decision not to include provision to offset incurred expenses has led to millions of dollars of wasteful spending".

"The now former MCPS Superintendent, Dr. Monifa McKnight, originally claimed that electric school buses would save the county money".

"We are going to be saving upwards of 6,500 gallons of diesel fuel per day," McKnight said during a 2022 press conference touting the initiative according to(https//electrek.co/2022/10/25/ largest-electric-school-bus-fleet-us/) to Electrek. Immediately this is going to cut cost by 50%". There is more to this story but I think the point has been made. Incidentally, there is no mention of the cost per kWh to charge the buses nor the cost of the necessary charging equipment.

A number of state school districts have gone the route of buying electric buses and they have experienced buses bursting into flame and many mechanical problems rendering the buses inoperative. In addition to the mechanical problems some of the purchasers of electric buses are left with inoperable buses while the manufacturer went bankrupt.

An article in Energy in Depth 10/14/24 – "Kamala Harris' electric school bus program is floundering, drawing comparison to 'border czar' work. Just the News. The EPA recently announced another $1 billion flowing into the Biden-Harris administration's effort to replace all diesel-powered school buses across the country with electric buses. This brings the total spent so far to nearly $3 billion out of the $5 billion program."

Another political misstep is in California where the current governor is trying to force refiners in the state to provide additional storage for product inventories in an effort to reduce price surges in the market. Additional inventory is not going to lower pump prices. The price of raw materials (feedstocks) and processing cost and government intervention such as this proposal are what generally impacts the price of fuel. Looking at California due to government controls you see higher electricity and natural gas costs that directly impact processing costs. Government regulations are very expensive. Forcing the construction of more tankage is not going to help. It adds to the overall refinery operating cost. In addition, the proposal just assumes the storage will just appear overnight. There is no consideration for the cost of additional tankage to accomplish that goal nor any consideration for the time necessary to go through the political gauntlet in order to get permits for new tankage (years). There is no consideration for the fact that there probably is no room for the tanks to be built on refinery property. Refineries operate with as little inventory as is possible to maintain production and their market demand. Excess inventory is expensive and an exposure to changes in market pricing.

Let's take a look at the so-called "Inflation Reduction Act". This from https://pinewsletter.com /gop-senator-exposes-biden/

Republican Sen., Shelly Moore Capito (R-WV) conducted an investigation on one of the grants given out by the EPA. The federal agency gave $50 million to a group called "Climate Justice Alliance." "Sounds like this group fights to protect the environment, right? Not quite."

From the post Millennial:

"U.S. Senator Shelley Moore Capito (R-WV) announced this week that the Environmental Protection Agency awarded a $50 million grant to Climate Justice Alliance using funds from the (so called) "Inflation Reduction Act" which democrats have touted as the biggest climate bill ever. Climate Justice Alliance "engages in pro-Hamas, anti-Israel, and anti-Semitic Activities," Capito said. "Time and time again when EPA officials come before our committee, they say" if only we had more money, more resources, more staff."

"This is what they're spending on? Many of us warned of the waste, fraud, and abuse. No one could have dreamed it would be this bad."

Sen. Moore showed images of the groups' website, which included the bulldozer that was used by Hamas in the horrific Oct.7th attack against Jewish civilians. The group also embraces the slogan, "Decolonize Palestine" something widely spread by radical anti-Israel groups."

"But it gets worse than that. It looks like this group, which just got $50 million from you, embraces most of the extremist views pushed by progressives. According to Sen. Capito, they support defunding the police and defunding the military. The group's affiliates also support initiatives the senator calls anti-American and anti- Semitic."

Your tax dollars at work! Just how did this $50 million expenditure reduce inflation? More government spending means more inflation. Once again, the EPA is a very dangerous government agency. Many of the EPA so called managers should be in jail.

As with the last chapter, this chapter could likely go on for several thousand pages as the U.S. government is the single most wasteful entity on the planet. In the interest of brevity, I'll stop here.

Chapter 9

Tax Payers Exposure to Others Owning EV"s

In a previous chapter some of the added personal costs associated with owning an EV were listed. This chapter will identify cost that all taxpayers experience even if they don't own an EV.

An Article found at https:patriot.com/study-the-true-costs-of-electric-vehicles-is-no-cost-savings/ "A thorough review of all costs of electric vehicles shows that they really don't offer the cost savings that EV pushers claim".

"Brent Bennet and Jason Isaac of the Texas Public Policy Foundation recently took the time to crunch the numbers and found that EV's are far more problematic than anyone thought".

EV zealots claim the vehicles cost less to drive, less to make, and less to maintain. But it appears they are not exactly telling the truth, according to the Daily Wire (https://www.dailywire.com/news/cost-of-driving-electric-vehicle-equal-to-paying-$17.33-per-gallonof-gasoline-study-finds).

"The authors of the study noted that no one has attempted to calculate the full financial benefit of the wide array of direct subsidies, regulatory credits, and subsidized infrastructure that contribute to the economic viability of EV's".

They go on to write "the average model year (MY) 2021 EV would cost $48,698 more to own over a 10 year period without $22 billion in government favors given to EV manufacturers and owners. Adding the costs of the subsidies to the true cost of fueling an EV would equate to an EV owner paying $17.33 per gallon of gasoline. And these estimates do not include the hundreds of billions more in subsidies in the Inflation Reduction Act (2022) for various aspects of the EV supply chain, particularly for battery manufacturing". All of these subsidies are your tax dollars being doled out to companies to produce a vehicle that they claim is environmentally superior to internal combustion engine vehicles. Not So!

The numbers above also don't include the unscheduled maintenance that EV's seem to be prone to, estimated at three times higher than an internal combustion engine vehicle. The uneducated purchasers of these cars are believing the government when they say that they are saving the planet. Nothing is further from the truth. And if in fact they are cheaper to make (as stated above), why do they cost the consumer 40% more to purchase. Many of these costs go directly to the EV buyer the rest are covered by the taxpayer that does not own an EV and is certainly not benefiting in any way from the vehicles.

"Nearly $22 billion in federal and state subsidies and regulatory credits suppressed the retail price of EV's in 2021 by an average of almost $50,000" The authors note "Car lots are swelling with unsold EV"s and the Ford Motor Company is losing over $70,000 on each EV it currently sells". Those losses are absorbed by Ford's stockholders.

"When we pay for a gallon of gasoline, we are paying for the entire infrastructure to refine, transport, and market that gasoline", they point out. "When an EV owner connects to the electric grid, how much are they paying for the extra generation, transmission and distribution costs that they are imposing on the grid, and will those embedded costs rise over time?" No doubt they will but not just for the EV owner those grid costs will be imposed on all end users. Your electric bill is going to increase even if you don't own an EV, as will your insurance rates for both car and home.

"The stark reality for proponents of EVs and for the dreamers in the federal government, who are using fuel economy regulations to force manufacturers to produce ever more EVs." What they conclude," is that the true cost of an EV is in no way close to the comparable ICEV (internal combustion engine vehicle). "Our conservative estimate is that the average EV accrues $48,698 in subsidies and $4,569 in extra charging and electricity costs over a 10-year period, for a total cost of $53,267 ,or $16.12 per equivalent gallon of gasoline. Without increased and sustained government favors, EVs will remain more expensive than ICEVs for many years to come." That $16.12 per gallon is not being paid by the EV owner, it is being paid by all U.S taxpayers just so an ignorant EV owner can walk around with his chest puffed out claiming he's saving the planet. They also completely discount the pollution that occurs where the rubber meets the road and stays there.

Another article in the Business Insider describes the cost of charging an EV at home or on the road compared to refueling an ICEV (internal combustion engine vehicle). "For trucks it roughly costs the same to fuel and charge. Whether you are fueling a Ford F series, Ram1500, or Chevrolet Silverado, it costs about $17.58 for every 100 miles. Trucks that take diesel cost about $17.10 to fuel for the same mileage. If you have a Rivan RIT or GMC Hummer, the cost to charge at home isn't much different. It's about $17.70 per 100 miles. The cost to charge, however, skyrockets 48% to about $26.38 if a driver is mostly charging in public." It's not clear if the numbers above include the 8% to 13% loss while charging due to heat generated by the process.

"Charging any of these vehicles at home will run about $12.62 per 100 miles while fueling one at gas station may run about $11.08. The total difference, while not incredibly substantial, may add up and when these drivers go to a public charging station they are likely to pay $16.10 per 100 miles." These are very general statements. I would guess they are calculated based on the national average for gasoline prices and the price per kWh. If you were to calculate the charging cost based on California price per kWh which is more than twice that of the national average the numbers would be much different, about 17% higher than twice the national average in some areas. The gasoline number would also look different though the California price per gallon of gasoline is higher than the national average it is not two times that average as is the kWh price. Therefore, an EV in California is going to cost considerably more to charge than elsewhere in the country. This does not take into account California commercial charging stations which would be considerably more than a home based charger. California tax payers will subsidize those higher cost owning an EV or not.

A few things pushers of EV's neglect to inform buyers is that if you get in even a minor fender bender you are likely to total the car. Once a battery pack is damaged it becomes useless. An additional "cost exposure" that is generally kept quiet is the trade in/ resale value of a used EV. Buyer beware, when you decide you want to buy a new car the trade in value for your used EV will be next to nothing. You need to keep the EV for at least 10 years for it to pay out in terms of cost saving relative to an ICEV, that of course assumes that electricity rates remain static (good luck with that). If you try to sell or trade the EV in before that you will take a substantial loss. And after 10 years you might as well park it on your front lawn and turn it into a flower bed because you won't be able to sell it. You might even be subject to hazardous waste fees. Trading the car in will have a very low value as car dealerships will have an extremely difficult time trying to sell a used EV. A dealership will likely not touch a 10 year old EV. The value would be negative as the disposal fees would be much more than the car is worth.

Americanactionnews.com posted an article: "It turns out The Biden-Harris EV push at massive cost-THOUSANDS OF BLUE COLLAR JOBS."

"There are fewer moving parts in an EV so it takes less manpower to produce an EV. Therefore, it requires fewer employees to build the car. There are currently about 120,000 Auto Workers Union jobs that will be eliminated if the manufactures go all EV. Having said that those jobs are taxpayers, once the job is gone so goes the tax revenues. To compound that issue the battery packs used in most EV's are made in China. So, we eliminate good tax paying jobs in the U.S to increase the trade deficit with China for very expensive car parts. These actions could literally destroy the American car manufacturing industry.

In addition to auto industry jobs, the energy sector stands to lose hundreds of thousands of jobs in the fields of production, manufacturing, distribution and retail. This applies not just to individual employees but to corporations as well. These mandates could bankrupt entire corporations that help maintain our economy. This translates to billions of dollars in tax revenues for, Federal, State and local government. The skill sets for those jobs in both industries, Auto and Energy, are not common to other industries, therefore you are looking at increasing the unemployment rolls by an enormous amount. This is what the Biden administration calls progress! Once the jobs are lost and the companies bankrupt, where will the tax dollars come from to support all of these new programs. It will be on the backs of the remaining working public until the government finds a way to do away with their jobs as well.

Something to keep in mind about these so-called scientists that all agree on global warming. Most of them are not climate scientists; they come from a variety of fields not remotely associated with climate studies. They just jump on board with the activists and try to promote the "Cause". That is the reason you hear "SCIENTISTS SAY" with no names or organizations mentioned when making their apocalyptic claims. When someone prefaces their catastrophic prediction with

"Scientist Say", without offering credentials, rest assured they are promoting a bogus claim and likely for monetary gain.

Since we have now crossed the threshold to a new administration one would hope we will get back to a common sense approach to policy making. The overbearing, stifling approach we have experienced for the last four years that gives bureaucrats free reign to control industry and the population must end. The so-called Environmental Protection Agency has done more damage to this country than any other organization in the history of the United States.

Authors Note

This book was an undertaking that has required much time and effort in research and fact checking. There has been a great deal of speculation relative to the impact of the use of oil and gas with no consideration for what those commodities bring to our everyday lives. Our country would not be what it is today and our citizens as prosperous as they are today if not for Petroleum and the companies that bring it to us.

You likely have ascertained, much of what is written is focused on the Biden / Harris administration's war on the American people, the Oil and Gas Industry, the Automobile Industry as well as the overall U.S Economy. The so-called "Inflation Reduction Act" is nothing but the "Green New Deal" in wolves' clothing. It has nothing to do with reducing inflation, it in fact causes inflation as it requires indiscriminate, wasteful government spending of billions of taxpayer funds. Those dollars are usually earmarked for anti-oil initiatives.

Through Executive Orders the Biden administration has all but crippled the Oil and Gas industry and propagated falsehoods using "climate change" as justification in the process. California has put in place many anti-oil initiatives so much so that at least one refiner is shutting down an LA based refinery while other refiners are contemplating similar action. Both Biden and Newsom have intentionally put countless road blocks in place in an attempt to sabotage the incoming Trump administration. Their intent was to make everything more difficult for

Trump policies to be enacted. Fortunately, President Trump and J.D. Vance and the people the president has surrounded himself with are far more intelligent than the former senile white house dweller/beach goer and the "Word Salad Blender" that preceded them.

We are now moving into a new era of common sense in government where the knee jerk reaction of environmental extremists will not force a knee jerk reaction to the so-called "Climate Change". Especially when the focus is on the United States compliance to global opinion/demands while China, Russia, and some United Nations members completely ignore what they preach. A prime example is the Paris Climate Agreement that gives China a pass on emissions compliance, until 2030, relative to what they are requiring in their mandate to the U.S. Meanwhile China gets to build 100 coal fired power plants per year. Do you think that China will somehow reverse their carbon footprint in 2030. They will completely ignore it and laugh all the way to the next Agreement meeting!

We will now be on a level playing field with the interest of the American people being the focus of future legislation. Hopefully, moving forward, we will have the opportunity to remove dangerous government agencies made up of bureaucrats with agendas that are detrimental to U.S. industries, the economy and to the people overall. One would hope we can get voter I.D to ensure true and fair elections across the country (that could even flip California) and maybe a line item veto to cut the ever present "Pork" from legitimate legislation as legislators pile-on a bill for self benefit.

My hope is that people who read what is written in the preceding pages will have a better understanding of the climate and what causes changes, the absolute necessity for oil and gas in their everyday lives and what is required to produce an Electric Vehicle. EV's are not environmentally friendly and providing the necessary minerals and infrastructure to support the conversion of the U.S. fleet is an absolute impossibility.